# SpringerBriefs in Business



For further volumes:
http://www.springer.com/series/8860

Caspar G. Chorus

# Random Regret-based Discrete Choice Modeling

## A Tutorial

Springer

Caspar G. Chorus
Department of Technology, Policy, Management
Delft University of Technology
Jaffalaan 5
2628 BX, Delft
The Netherlands

ISSN 2191-5482 e-ISSN 2191-5490
ISBN 978-3-642-29150-0 e-ISBN 978-3-642-29151-7
DOI 10.1007/978-3-642-29151-7
Springer Heidelberg New York Dordrecht London

Library of Congress Control Number: 2012935801



Printed on acid-free paper

Springer is part of Springer Science+Business Media (www.springer.com)

# Preface

This book presents a hands-on introduction to a new discrete choice modeling approach based on the behavioral notion of regret-minimization. This so-called Random Regret Minimization-approach (RRM) forms a counterpart of the Nobel prize winning Random Utility Maximization-approach (RUM) to discrete choice modeling, which has for decades dominated the field of choice modeling and adjacent fields such as transportation, marketing, and environmental economics. Being as parsimonious as conventional RUM-models and compatible with popular software packages, the RRM-approach provides an alternative and appealing account of choice behavior. A variety of empirically well-established behavioral phenomena that are not captured by conventional RUM-models readily emerge from the RRM-model's structure in a way that is consistent with the model's underlying behavioral premises. Since its introduction in 2010, the RRM-approach has been used by a growing number of leading choice modelers in a variety of decision-making contexts, showing a promising potential in terms of model fit and predictive ability. Rather than providing highly technical discussions as usually encountered in scholarly journals covering discrete choice models, this book aims to allow readers to explore the RRM-approach and its potential and limitations hands-on, based on a detailed discussion of examples. This book is written for students, scholars, and practitioners who have a basic background in choice modeling in general and RUM-modeling in particular. It has been taken care that all concepts and results should be clear to readers who do not have an advanced knowledge of econometrics. Following a brief introduction, Chap. 2 presents the RRM-approach to discrete choice modeling. Comparisons with the RUM-based Multinomial Logit-model are provided where relevant. In Chap. 3, the focus is on how to estimate RRM-models and how to interpret estimation results and use them for forecasting—the Chapter discusses data-requirements and software issues, and presents an empirical example in-depth. In Chap. 4, the general applicability of the RRM-model is discussed, and its strong points and limitations are highlighted. Chapter 5 presents a selection of recent developments in RRM-modeling.

# Contents

1 **Introduction** . . . . . . . . . . 1
References . . . . . . . . . . 4

2 **A Random Regret Minimization-based Discrete Choice Model** . . . 5
2.1 A Random Regret-Function . . . . . . . . . . 5
2.2 A Comparison with (Linear-Additive) Utility Maximization . . . . 9
2.3 Regret-based Choice Probabilities and a RRM-based MNL-Model . . . . . . . . . . 13
References . . . . . . . . . . 14

3 **Empirical Application of Random Regret Minimization-Models** . . 17
3.1 The Data . . . . . . . . . . 17
3.2 Model Estimation . . . . . . . . . . 19
3.3 Model Fit . . . . . . . . . . 21
3.4 Interpretation of Parameters . . . . . . . . . . 23
3.5 Market Share Forecasting . . . . . . . . . . 29
3.6 Out-of-Sample Validity . . . . . . . . . . 32
References . . . . . . . . . . 34

4 **Applicability of Random Regret Minimization-Models, and Their Strong and Weak Points** . . . . . . . . . . 35
4.1 General Applicability . . . . . . . . . . 35
4.1.1 Non-Generic Attributes . . . . . . . . . . 35
4.1.2 Constants . . . . . . . . . . 36
4.1.3 Non-continuous (Categorical) Attributes . . . . . . . . . . 37
4.1.4 Hybrid RUM-RRM Models . . . . . . . . . . 38
4.1.5 Interactions with Socio-Demographic Variables . . . . . . . 38
4.2 RRM: Strong Points . . . . . . . . . . 38
4.3 RRM: Weak Points . . . . . . . . . . 39
References . . . . . . . . . . 41

**5 Selection of Recent Developments in RRM-Modeling** . . . . . . . . . . 43
5.1 Random Regret Minimization for Decision-Making Under Uncertainty . . . . . . . . . . . . . . . . . . . . . . . . . . . . . . . . 43
5.2 RRM for Modeling Household or Group Decision-Making . . . . . 46
5.3 The Expected Regret of a Choice-Situation: An RRM-based Logsum . . . . . . . . . . . . . . . . . . . . . . . . . . . . 48
5.4 An RRM-based Formulation of 'Value of Time' . . . . . . . . . . . . 50
References . . . . . . . . . . . . . . . . . . . . . . . . . . . . . . . . . . . . . 52

**Appendix: Investigating the Assumption of i.i.d. Errors in RRM-Models** . . . . . . . . . . . . . . . . . . . . . . . . . . . . 53

# Chapter 1
# Introduction

Since their inception in the 1970s (McFadden 1974), Discrete Choice Models (DCMs from here on) have been used to explain choice-behavior of individuals and predict market shares for products and services in a wide variety of contexts. Numerous studies worldwide have been performed over the years to provide quantitative analyses and forecasts of choice behavior and market shares in fields as diverse as transportation, health care, consumer choice, and environmental economics—to name a few. DCMs use observed choices between different options (for example: stated or revealed choices between different travel modes on a given route) to derive the underlying preferences of individuals. When one has information about the characteristics of the different choice options—such as the travel times and costs of the different travel modes—DCMs make it possible to estimate the weights that decision-makers attach to these different characteristics when making decisions. Knowing these weights is valuable in itself. However, they can also be put to use to predict the effect on market shares of, say, an increase in bus-fare of a particular magnitude. This ability of DCMs to provide quantitative estimates of decision-makers' preferences, and of the market shares that are the aggregate result of these preferences, has made them very popular among scholars and practitioners alike

Basically, each DCM is defined in terms of a combination of an assumed decision rule and an assumed error term structure. The error terms that enter DCMs represent the fact that the researcher cannot 'look into the heads' of decision-makers, so that part of what drives their choice-behavior remains unobserved. Lately, the DCM-field has seen numerous innovations in error term specifications, allowing for an ever more realistic representation of the unobserved part of behavior. This tutorial will not pay much attention to these developments; note that excellent textbooks exist that cover them in much detail (e.g., Train 2003). Instead, the tutorial will mainly focus on the *decision rule* assumed in DCMs.

In terms of the assumed decision-rule, DCMs, like every model, present a simplified version of reality: although it is well known that choice-behavior may

C. G. Chorus, *Random Regret-based Discrete Choice Modeling*,
SpringerBriefs in Business, DOI: 10.1007/978-3-642-29151-7_1,

be the result of numerous interrelated processes and impulses, a DCM must reduce this complexity to a great extent in order to be of practical use. One of the ways in which DCMs reduce the complexity of actual behavior is by assuming a relatively straightforward decision-rule which translates decision makers' preferences and tastes (e.g., a dislike of long travel times), in combination with the characteristics of choice-options (e.g., a bus service's travel time), into predicted choice patterns (e.g., a particular bus service's market share on a given route).

More specifically, the overwhelming majority of DCMs are based on a so-called utility maximization-based decision rule. This decision rule postulates that when choosing between different options, decision makers attach a utility to each option and choose the one that has the highest utility. Apart from an error term, this utility usually consists of a linear-additive function of characteristics of the choice-option and associated parameters (the latter represent decision weights associated with these characteristics). DCMs that are based on such utility maximization-based decision rules are called Random Utility Maximization-models (the term 'Random' stands for the error term that is added). This utilitarian category of DCMs (from here on: RUM-models or RUM-based DCMs[1]) has been by far the most used DCM since its inception, earning the main developer (Daniel McFadden) a Nobel Prize. The popularity of RUM-models both in- and outside academia is a direct result of their tractability, ease of use and solid foundation in microeconomic axioms.

Notwithstanding this popularity of the utility-based modeling approach, interest in the incorporation of alternative decision rules in DCMs has existed for as long as RUM-models have. This tutorial focuses on such a non-utilitarian DCM called Random Regret Minimization or RRM, which has been recently proposed by the author of this tutorial (Chorus 2010).[2] This regret-based DCM-approach is based on the notion that when choosing, people anticipate and aim to minimize regret rather than maximize utility. Regret arises when one or more non-chosen alternatives perform better than the chosen one in terms of one or more attributes. The notion that people are regret-minimizers has been very well established empirically in a variety of fields (e.g., Zeelenberg and Pieters 2007), but the translation of this behavioral notion in an operational DCM that enables the econometric analysis of multi-attribute alternatives in multi-alternative choice sets is new. RRM distinguishes itself from other extensions of and alternatives for RUM-models by presenting a tractable, parsimonious model form that is easily estimable using

[1] Unless explicitly stated otherwise, a reference to RUM-models in this tutorial implies a reference to linear-additive RUM-based multinomial logit (or MNL) models as presented in Sect. 2.2.

[2] Note that an earlier version of the RRM-model has been presented in (Chorus et al. 2008). However, the RRM-model form presented in that paper featured a non-smooth likelihood-function which hampered the model's applicability and usability (for example: it relied on handwritten code). This manual focuses on a more recent model form, published in 2010 (Chorus 2010); this latter model form features a smooth likelihood-function and can be estimated using standard software-packages.

standard software packages. In other words: the model consumes no more parameters than conventional (linear-additive) RUM-based DCMs, and it is user-friendly. Notwithstanding that the model has only been introduced very recently, it has already been successfully applied by a rapidly growing number of researchers from a variety of leading research groups worldwide.[3] Applications include but are not limited to travelers' choices between destinations, travel modes, routes, departure times, parking lots, and travel information services; tourists' choices between recreational activities; patients' choices between medical treatments; consumers' choices between vehicle-types; politicians' choices between policy-options; and singles' choices between on-line dating profiles.

Motivated by the growing popularity of RRM-based DCMs, this tutorial aims to provide an introduction to regret-based choice modeling for students and practitioners, as well as for scholars with a basic understanding of discrete choice-modeling and an interest in RRM. In contrast with published scholarly papers on RRM, whose main aim is to communicate new research findings to fellow discrete choice-modelers, the tutorial has an educational purpose. It aims to help the reader (i) understand the RRM-approach to discrete choice modeling (ii) appreciate how RRM-models differ from RUM-models conceptually and operationally, (iii) understand how the RRM-model can be used in practice to analyze choice-data, and (iv) appreciate its potential and limitations. Although equations are presented wherever relevant, the manual is written in a way that facilitates understanding of RRM-models and their properties also for readers that have only moderate experience in working with mathematical formulations. However, a working knowledge of discrete choice modeling in general and the RUM-based MNL-model is required, as is a basic understanding of stated choice-data collection methods.[4] The majority of examples used is obtained from the field of transportation in general and travel behavior in particular. However, applicability of derivations, results and discussions in other fields than transportation will be obvious and intuitive.

Chapter 2 presents the RRM-approach to discrete choice modeling. Comparisons with the RUM-based MNL-model are provided where relevant. In Chap. 3, the focus is on how to estimate RRM-models and how to use estimation results for forecasting—it discusses data-requirements, software issues and presents an empirical example in-depth. In Chap. 4, the general applicability of the RRM-model is discussed, and its strong points and limitations are discussed. Chapter 5 presents a selection of recent developments in RRM-modeling.

I would like to thank Eric Molin for reading and commenting on an earlier version of the first four chapters of this tutorial. Furthermore, I have had many very

[3] See Chorus (2012) for a recent overview of empirical applications.

[4] See Ben-Akiva and Lerman (1985) and Train (2003) for examples of textbooks that provide excellent introductions to discrete choice-modeling and RUM's MNL-model (besides offering much material for more advanced choice-modelers as well). Hensher et al. (2005) provide an excellent introduction to stated choice methods, as well as much material at a more advanced level.

fruitful discussions about RRM with a variety of choice modelers, especially with people with whom I have been or currently am doing research and writing papers on this topic: Jan Anne Annema, Theo Arentze, Richard Batley, Matthew Beck, Shlomo Bekhor, Esther de Bekker-Grob, Michel Bierlaire, Michiel Bliemer, Marco Boeri, Bill Greene, David Hensher, Stephane Hess, Anco Hoen, Gerard de Jong, Mark Koetse, Niek Mouter, John Nellthorp, John Rose, Ric Scarpa, Mara Thiene, Harry Timmermans, Tomer Toledo, and Bert van Wee. Discussions with these scholars have refined my own thinking on RRM, and I would like to thank them for that. Support from the Netherlands Organization for Scientific Research (NWO), in the form of VENI-grant 451-10-001, is gratefully acknowledged.

## References

Ben-Akiva ME, Lerman SR (1985) Discrete choice analysis: theory and application to travel demand. The MIT Press, Cambridge

Chorus CG, Arentze TA ,Timmermans HJP (2008) A random regret minimization model of travel choice. Transp Res B 42:1−18

Chorus CG (2010) A new model of random regret minimization. Eur J Transp Infrastruct 10:181–196

Chorus CG (2012) Random regret minimization: an overview of model properties and empirical evidence. Transp Rev 32:75–92

Hensher DA, Rose JM, Greene WH (2005) Applied choice analysis: a primer. Cambridge University Press, Cambridge

McFadden D (1974) Conditional logit analysis of qualitative choice-behavior. In: Zarembka P (ed) Frontiers in econometrics. Academic Press, New York

Train KE (2003) Discrete choice methods with simulations. Cambridge University Press, Cambridge

Zeelenberg M, Pieters R (2007) A theory of regret regulation 1.0. J Consum Psychol 17:3–18

# Chapter 2
# A Random Regret Minimization-based Discrete Choice Model

**Abstract** This Chapter presents the RRM-model. First, the Random Regret-function is presented and explained (Sect. 2.1). Subsequently, this function is compared with the classical linear-additive Random Utility-function (Sect. 2.2). Finally, it is shown how the Random Regret-function translates into MNL-type choice probabilities for a particular distribution of the random error terms (Sect. 2.3).

## 2.1 A Random Regret-Function

The key behavioral notion on which the RRM-model is built is that people, when choosing, compare a considered alternative with each of the other available alternatives in terms of each characteristic (or from here on: attribute), and that they wish to avoid the situation where a chosen alternative is outperformed by one or more other alternatives on one or more attributes (which would cause regret).[1] Importantly, in contrast with other models and theories that are based on regret-minimization, the RRM-model postulates that anticipated regret is also a determinant of choices when there is no uncertainty about the performance of alternatives. The RRM-model postulates that as long as alternatives are characterized in terms of multiple attributes,

[1] Obviously, the level of anticipated regret that is associated with a particular alternative will vary between individuals. More specifically, different individuals may have different tastes and perceptions regarding alternatives and their attributes. Mathematically, this heterogeneity across individuals can be expressed by making relevant terms in the regret equation presented below (such as $\beta$ and $\varepsilon$) individual-specific by means of an index (usually '$n$'). In this tutorial, for reasons of readability, no such indices are used. As a result, equations refer to (the tastes and perceptions of) an average or 'representative' individual.

C. G. Chorus, *Random Regret-based Discrete Choice Modeling*,
SpringerBriefs in Business, DOI: 10.1007/978-3-642-29151-7_2,

which implies that trade-offs have to be made by the decision-maker, there will be regret in the sense that there will generally be at least one non-chosen alternative that outperforms a chosen one in terms of one or more attributes.

More specifically, the RRM-model is designed to incorporate the following seven behavioral intuitions relating to the anticipated regret associated with a considered alternative:

1. when a considered alternative *outperforms* another alternative in terms of a particular attribute, the comparison of the considered alternative with the other alternative on that attribute does not generate anticipated regret.
2. When a considered alternative *is outperformed* by another alternative in terms of a particular attribute, the comparison of the considered alternative with the other alternative on that attribute generates anticipated regret.
3. Anticipated regret increases with the importance of the attribute on which a considered alternative is outperformed by another alternative.
4. Anticipated regret increases with the magnitude of the extent to which a considered alternative is outperformed by another alternative on a particular attribute.
5. Anticipated regret increases with the number of attributes on which the considered alternative is outperformed by another alternative.
6. Anticipated regret increases with the number of alternatives that outperform a considered one on a particular attribute.
7. Anticipated regret is, from the perspective of the analyst, partially 'observable' (in the sense that it can be explicitly linked to observed variables) and partially 'unobservable'.

The following equation (Eq. 2.1) gives a formulation of regret that is consistent with these intuitions:

$$RR_i = R_i + \varepsilon_i = \sum_{j \neq i} \sum_m \ln\left(1 + \exp\left[\beta_m \cdot \left(x_{jm} - x_{im}\right)\right]\right) + \varepsilon_i \tag{2.1}$$

$RR_i$ denotes the random (or: total) regret associated with a considered alternative $i$

$R_i$ denotes the 'observed' regret associated with $i$

$\varepsilon_i$ denotes the 'unobserved' regret associated with $i$

$\beta_m$ denotes the estimable parameter associated with attribute $x_m$

$x_{im}$, $x_{jm}$ denote the values associated with attribute $x_m$ for, respectively, the considered alternative $i$ and another alternative $j$.

Before discussing this function in more depth, it should be noted that of course, constants can be added to regret-functions, to represent the mean unobserved regrets associated with particular alternatives. Also, note that attributes may take the form of continuous variables as well as variables of categorical measurement level. Furthermore note that socio-demographic variables (such as age, gender, income and education level) may enter the regret-function to express segmentations of the population in terms of preferences for alternatives and tastes for attributes. Finally,

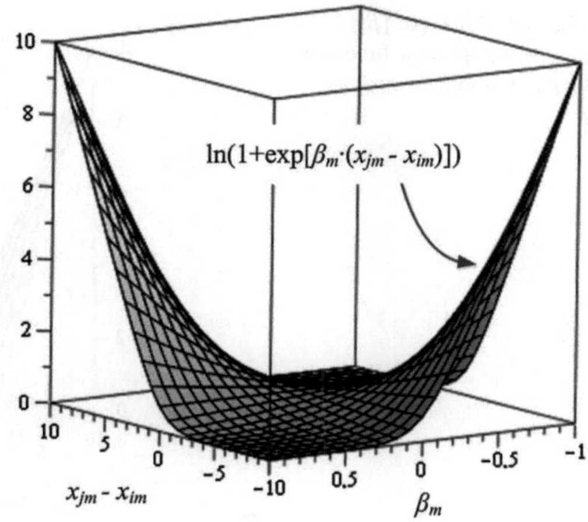


**Fig. 2.1** Attribute-level regret as a function of $\beta_m$ and $(x_{jm} - x_{im})$

note that in this chapter the focus is on attributes that are common, or shared, or generic, across alternatives. See Chap. 4 for an in-depth discussion of how the RRM-model deals with constants, interactions with socio-demographic variables, non-continuous variables, and with attributes that are specific to particular alternatives.

Returning to the regret-equation presented above: the term $\ln(1 + \exp[\beta_m \cdot (x_{jm} - x_{im})])$ is the core of this equation: it forms a measure of the amount of regret that is associated with comparing a considered alternative $i$ with another alternative $j$ in terms of a particular attribute $x_m$. This attribute-level regret is computed for each of the bilateral comparisons with other alternatives, and for all available attributes; the summation of these attribute-level regret terms (totaling $M * (J-1)$ terms in a situation where there are $M$ attributes and the choice set contains $J$ alternatives) forms the observed regret that is associated with the considered alternative. In light of the important role of attribute-level regret, also when it comes to deriving and interpreting the properties of the RRM-model, it is worth paying additional attention to this regret-kernel.

First, let us investigate whether the third and fourth behavioral intuitions formulated above (which refer to attribute-level regret) are captured in the used measure. As $\ln(1 + \exp[\beta_m \cdot (x_{jm} - x_{im})])$ is a monotonically increasing function of both $\beta_m$ and $(x_{jm} - x_{im})$ it is easily seen that attribute-level regret increases with the importance of the attribute on which a considered alternative is outperformed by another alternative, and with the magnitude of the extent to which the alternative is outperformed by another alternative on that attribute. Figure 2.1 provides a visual account of this argument, and illustrates how attribute-level regret emerges for the situation where higher attribute-values are preferred over lower ones (implied by a positive sign of $\beta_m$), respectively for the situation where lower attribute-values are preferred over higher ones (implied by a negative sign of $\beta_m$).

More specifically, when higher attribute-values are preferred over lower ones, regret emerges to the extent that $x_{jm}$ becomes larger than $x_{im}$ and to the extent that

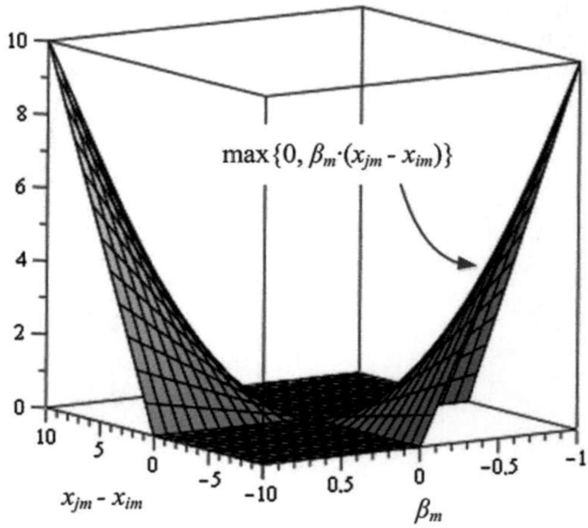


**Fig. 2.2** Max {0, [$\beta_m \cdot (x_{jm} - x_{im})$]} as a function of $\beta_m$ and $(x_{jm} - x_{im})$

$\beta_m$ becomes more positive. In contrast, when lower attribute-values are preferred over higher ones, regret emerges to the extent that $x_{jm}$ becomes smaller than $x_{im}$ and to the extent that $\beta_m$ becomes more negative. When the product $\beta_m \cdot (x_{jm} - x_{im})$ becomes negative (i.e., when the considered alternative outperforms the other alternative with which it is compared in terms of the attribute), attribute-regret starts to approach zero.[2]

Note that strictly speaking, the fact that attribute-regret *approaches* rather than *equals* zero when the considered alternative outperforms the other alternative implies that the first of the seven behavioral intuitions formulated at the beginning of this Chapter is violated. More generally speaking, although the measure $\ln(1 + \exp[\beta_m \cdot (x_{jm} - x_{im})])$ turns out to form a behaviorally intuitive measure of attribute-level regret, it seems a bit far-fetched at first sight. Indeed, a more intuitive and easy to interpret measure of attribute-level regret exists, in the form of max $\{0, [\beta_m \cdot (x_{jm} - x_{im})]\}$. This measure, introduced in Chorus et al. (2008),[3] actually represents attribute-level regret in a more direct way than does $\ln(1 + \exp[\beta_m \cdot (x_{jm} - x_{im})])$: when a considered alternative outperforms another alternative in terms of the attribute, regret equals zero (hence, the first behavioral intuition is not violated by this measure of attribute-regret). When the considered alternative is outperformed, regret equals the product of the importance of the attribute and the difference in attribute-values. Figure 2.2 plots max$\{0, [\beta_m \cdot (x_{jm} - x_{im})]\}$ as a function of $\beta_m$ and $(x_{jm} - x_{im})$, thus forming a counterpart of Fig. 2.1. The resemblance with Fig. 2.1 is very obvious: it is easily seen that $\ln(1 + \exp[\beta_m \cdot (x_{jm} - x_{im})])$

---

[2] It should be noted at this point that during the estimation process the sign of parameters is estimated together with their magnitude. That is, no a priori expectations need to be formulated by the analyst in terms of whether higher attribute-values are preferred by the decision-maker over lower ones, or vice versa.

[3] See Chorus et al. (2008, 2009) and Hess et al. (in press) for applications of this particular form of the RRM-model.

forms a close approximation of $\max\{0, [\beta_m \cdot (x_{jm} - x_{im})]\}$, especially when the absolute values of $\beta_m$ and/or $(x_{jm} - x_{im})$ become larger.

The main difference between the two measures is that the former function is smooth while the latter (the one with the max-operator) is not. It turns out that it is exactly this 'non-smoothness' of $\max\{0, [\beta_m \cdot (x_{jm} - x_{im})]\}$ what makes this measure a less useful one for discrete choice modeling. More specifically, the fact that the function is discontinuous around zero (see Fig. 2.2) makes that the partial derivatives of the function with respect to $\beta$s and $x$s cannot be computed in that area. This causes non-trivial theoretical and practical difficulties in the process of Maximum Likelihood-based model estimation as well as in the process of computing elasticities. This non-smoothness also results in practical difficulties in the sense that conventional software-packages do not support this kind of non-smooth functions and hence the researcher has to rely on handwritten code for model estimation. This obviously greatly hampers the model's usability, especially among students and practitioners. These theoretical and practical issues are solved by adopting the smooth function plotted in Fig. 2.1, which was introduced in Chorus (2010) and which is used in the remainder of this tutorial.[4]

## 2.2 A Comparison with (Linear-Additive) Utility Maximization

Before deriving RRM-choice probability-formulations for the random regret-function presented in Sect. 2.1, it is instructive to compare the random regret-function with the random utility function that has dominated the field of DCM for decades. More specifically, the random regret-function is contrasted with its most natural counterpart: the so-called *linear-additive* random utility-function. The term 'linear-additive' refers to the fact that observed utility is a summation of terms that each consist of a product of a parameter and an attribute-value. It is this function that utility-maximizers aim to maximize by choosing between different alternatives. In notation (Eq. 2.2):

$$U_i = V_i + \varepsilon_i = \sum_m \beta_m \cdot x_{im} + \varepsilon_i \tag{2.2}$$

$U_i$ denotes the random (or: total) utility associated with a considered alternative $i$
$V_i$ denotes the 'observed' utility associated with $i$
$\varepsilon_i$ denotes the 'unobserved' utility associated with $i$
$\beta_m$ denotes the estimable parameter associated with attribute $x_m$
$x_{im}$ denotes the value associated with attribute $x_m$ for the considered alternative $i$.

[4] Note that, although the use of smooth attribute-regret-function instead of the non-smooth one is inspired mostly by pragmatic reasons as argued above, there is a deeper connection between the two functions as well. That is, when ignoring a constant, $\ln(1 + \exp[\beta_m \cdot (x_{jm} - x_{im})])$ gives the expectation of $\max\{0, [\beta_m \cdot (x_{jm} - x_{im})]\}$ when the two terms between curly brackets are considered i.i.d. random variables with Extreme Value Type I-distribution (having a variance of $\pi^2/6$). A reason for this stochasticity might be that the researcher is only able to assess these two terms up to a random error.

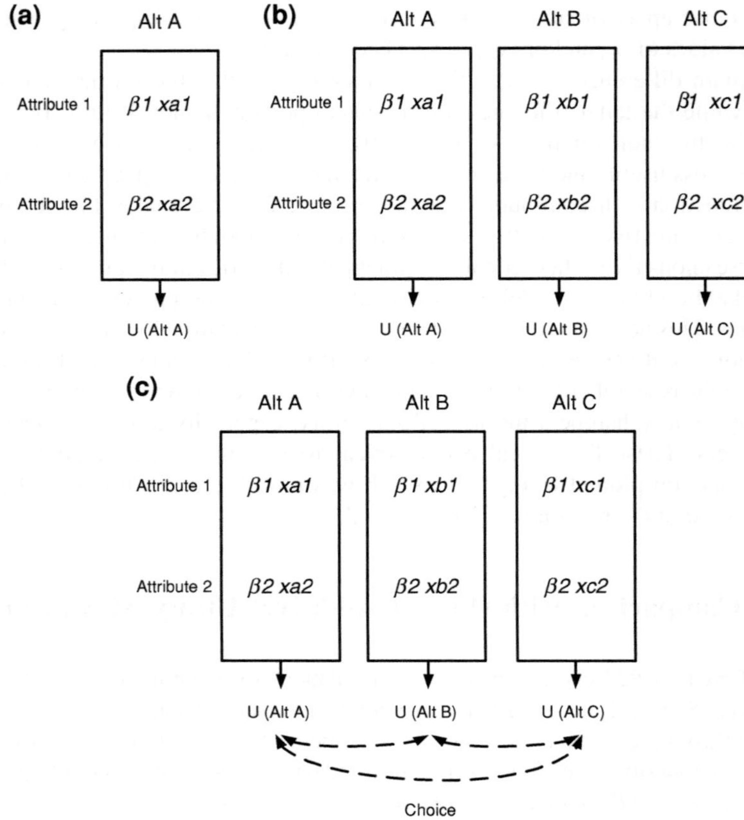


**Fig. 2.3** A linear-additive utility maximization-based decision process (*solid arrows* represent summations, *dashed arrows* represent comparisons)

The conceptual differences between the utility-function presented directly above and the regret-function presented in Sect. 2.1 can be understood by inspecting Fig. 2.3: the Figure depicts the decision-process assumed in linear-additive RUM-models,[5] in the context of the following example: a decision-maker chooses between three alternatives A, B, and C (say, a train, car and bus-mode),

[5] Note that strictly speaking, DCMs do not really assume particular *processes* (in the sense that they do not postulate a particular order of decision-making steps). Rather, the mathematical formulation of the linear-additive RUM- model is in fact consistent with a range of underlying decision processes. Nonetheless, throughout the literature the linear-additive RUM-model form is generally considered to be the mathematical representation of the decision process described and visualized on this and the next page. It is instructive at this point to assume this particular order in decision-making steps as it highlights the ways in which RRM- and RUM-based decision rules differ in a conceptual sense.

and alternatives are evaluated in terms of two attributes (x1 and x2, say, travel time and travel cost). Linear-additive RUM-models assume that before a choice is made, the utility of each alternative is computed. Figure 2.3a depicts this process for alternative A (the train mode).

The decision-maker is assumed to 'compute' the utility of the train alternative by means of combining ('multiplying') his or her tastes (or: decision-weights) with the two mode-specific attribute-values. In other words: his or her taste for travel time is combined with the train's travel time and the same is done for the travel cost-attribute. Subsequently, the two combinations (one for time and one for cost) are summed together to form a measure of train-utility.[6] This process is repeated for the car and bus mode (Fig. 2.3b). After having this way 'computed' utilities for each of the three mode-options, the three utilities are compared and the highest utility alternative is chosen (Fig. 2.3c).

The RRM-model is based on quite a different assumed decision-making process (see Fig. 2.4): first, all alternatives are bilaterally compared on attribute x1 (travel time). The decision-maker does this by means of combining his or her taste for (decision weight attached to) travel time with travel time differences between train and car, train and bus, and car and bus, respectively. This process is done both ways (so: the car-option is compared with the bus-option in terms of travel time, and the bus-option is also compared with the car-option in terms of travel time, etc.); this results in 3*2 = 6 attribute-regret terms. This process is repeated for the cost attribute (Fig. 2.4b), after which attribute level-regrets are summed together for each alternative (resulting in 3*2*2 = 12 attribute-regrets in total, or 2*2 = 4 attribute-regrets per alternative). This way, measures of overall regret are obtained for the train-, car- and bus-option; the regrets are compared and the mode with minimum regret is chosen.

In sum, in terms of the assumed underlying decision-process the (linear-additive) utility maximization-rule and the regret minimization-rule differ in the sense that, while the utility maximization-rule assumes that comparisons between alternatives are only made at the level of aggregated utilities, the regret minimization-rule postulates that comparisons between alternatives are made at the level of each attribute, as well as at the level of aggregated regrets. Phrased differently: whereas a utility maximizer is focused on the performance of a considered alternative itself (or: in isolation), a regret minimizer is focused on how a considered alternative compares with other alternatives in terms of every conceivable aspect.

However, it should at this point again be noted that to a considerable extent the differences highlighted above are artificial: strictly speaking, choice models do not assumed particular decision processes, and the same mathematical model-formulation is generally consistent with a range of underlying processes. The only

---

[6] Note that the random error is ignored in this example as it refers to the *analyst's* lack of knowledge and as such is irrelevant from an individual decision-maker's point of view.

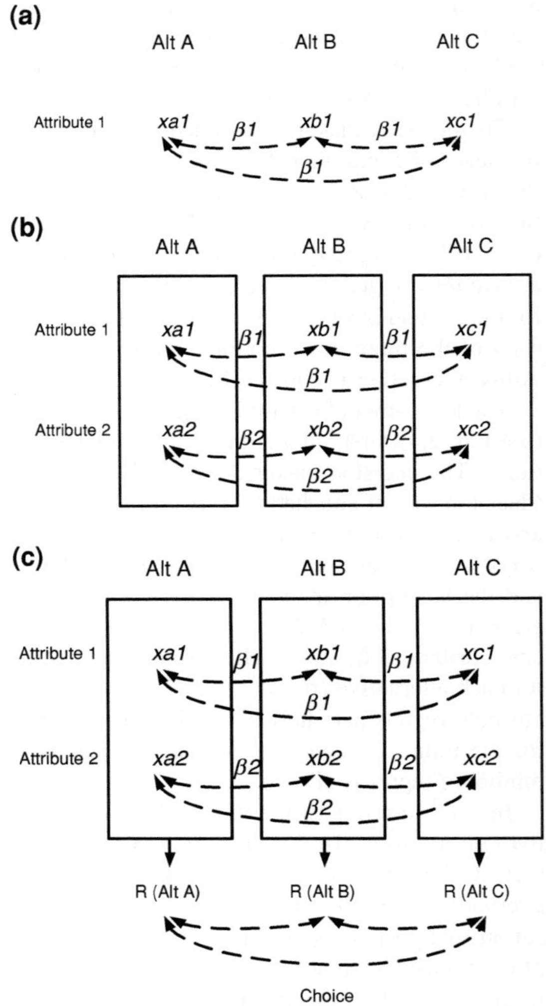


**Fig. 2.4** A regret minimization-based decision process (*solid arrows* represent summations, *dashed arrows* represent comparisons)

formal difference between the RRM-model and a linear-additive RUM-model is that in the former, attributes of other alternatives codetermine the utility (called regret) of a considered alternative, and that they do so in an asymmetric, non-linear way.

As a consequence of the slightly fuzzy nature of the conceptual differences between the two models in terms of their assumed decision processes, it is more important to discuss how the two models differ in terms of their predictions (i.e., in terms of the choice probabilities they assign to different alternatives). The next chapter will provide such a comparison. However, before this comparison can be made, choice probabilities have to be derived for the RRM-model.

## 2.3 Regret-based Choice Probabilities and a RRM-based MNL-Model

Having established and discussed in-depth the random regret-function (as well as how it contrasts conceptually with its natural RUM-counterpart, the linear-additive random utility-function), the next step is to present a regret-based choice probability-formulation which gives the probability that a regret-minimizer chooses a particular alternative from a choice set. Obviously, should a researcher know the total (or: random) regret associated with each alternative, then he or she can readily determine the chosen alternative (which is the one with minimum regret). However, since part of the regret that is associated with a particular alternative is 'unobserved' by the analyst as is represented by the random error term, he or she can only predict choices up to a probability. Like is the case for RUM-models, this formulation of this probability depends on the particular distribution assumed for the error terms. For RUM-models it has been found (McFadden 1974) that the most convenient (because: closed form) formulation of choice probabilities is obtained when errors are assumed to be i.i.d. Extreme Value Type I-distributed.[7] This result (leading to the RUM-based MNL-or Multinomial Logit-model of discrete choice) can be used to obtain the same kind of closed form-probability formulation for regret-minimizers. This is most easily understood by first briefly revisiting the derivation of RUM-choice probabilities. More specifically, the probability that a utility-maximizer chooses alternative $i$ from the set of $J$ available alternatives, given i.i.d. Extreme Value Type I-errors, can be put as follows (Eq. 2.3):

$$\begin{aligned} P(i) &= P\left(U_i > U_j, \forall j \neq i\right) \\ &= P\left(V_i + \varepsilon_i > V_j + \varepsilon_j, \forall j \neq i\right) \\ &= \frac{\exp(V_i)}{\sum_{j=1..J} \exp(V_j)} \end{aligned} \tag{2.3}$$

It can now be easily seen (Eq. 2.4) that in a regret-minimization setting, the assumption that *the negative of* the random errors is i.i.d. Extreme Value Type I-distributed[8] results in a very similar, MNL-formulation of choice probabilities:

---

[7] The term i.i.d. stands for identically and independently distributed. This means that errors assigned to different alternatives are uncorrelated, and are drawn from the same distribution (with the same variance). This variance is usually fixed to $\pi^2/6$, which indirectly implies a normalization of systematic utility. In this tutorial, the scale of the utility or regret is always normalized this way, and is therefore not explicitly mentioned in equations.

[8] See the Appendix for a discussion of the validity of the assumption of i.i.d. errors in the context of RRM-models.

$$\begin{aligned}P(i) &= P\left(RR_i < RR_j, \forall j \neq i\right)\\ &= P\left(-RR_i > -RR_j, \forall j \neq i\right)\\ &= P\left(-(R_i + \varepsilon_i) > -(R_j + \varepsilon_j), \forall j \neq i\right)\\ &= P\left(-R_i - \varepsilon_i > -R_j - \varepsilon_j, \forall j \neq i\right)\\ &= \frac{\exp(-R_i)}{\sum_{j=1..J} \exp(-R_j)}\end{aligned} \tag{2.4}$$

The fact that the RRM-model features MNL-type choice probabilities (in combination with the fact that it has a smooth regret-function) comes with many benefits. Particularly it implies that, although the underlying behavioral premises of the RRM-model are fundamentally different from those of conventional RUM-models (see Sect. 2.2), the RRM-based MNL-model can use many of the econometric tools contained in the very comprehensive and well-understood 'toolbox' that has been developed over the past three decades in the context of the RUM-based MNL-model. This includes the use of estimation routines embedded in standard software packages.

It should be noted that, although in the remainder of this tutorial the focus will be on the MNL-model form presented above, extension towards so-called Mixed Logit-model forms is straightforward. That is, by adding error terms or by allowing parameters or the scale factor to vary randomly across individuals, so-called nesting and panel effects as well as random taste- and/or scale-heterogeneity can be accommodated in RRM-based Mixed Logit models. Translation of RRM-based MNL-models towards RRM-based Mixed Logit models is equivalent to the translation of RUM-based MNL-models towards RUM-based Mixed Logit models and hence will not be covered in this tutorial (see for example Train (2003) for an excellent treatment of RUM-based Mixed Logit models).

Finally, it should be noted that in binary choice situations (containing only two alternatives), the RRM-based MNL and the RUM-based MNL result in the same choice probabilities. For the interested reader, a formal proof is provided in the appendix of Chorus (2010).

## References

Chorus CG (2010) A new model of random regret minimization. Eur J Transp Infrastruct 10:181–196

Chorus CG, Arentze TA, Timmermans HJP (2008) A random regret minimization model of travel choice. Transp Res B 42:1–18

Chorus CG, Arentze TA, Timmermans HJP (2009) Spatial choice: a matter of utility or regret? Environ Plan B 36:538–551

Hess S, Stathopoulos A, Daly A (in press) Mixing of behavioral processes: a modeling framework and three case-studies. Transportation. doi:10.1007/s11116-011-9365-6

McFadden D (1974) Conditional logit analysis of qualitative choice-behavior. In: Zarembka P (ed) Frontiers in econometrics. Academic Press, New York

Train KE (2003) Discrete choice methods with simulations. Cambridge University Press, Cambridge

# Chapter 3
# Empirical Application of Random Regret Minimization-Models

**Abstract** This chapter presents an in-depth discussion of how the RRM-based MNL-model is estimated, and how estimation results are interpreted and used for forecasting. This is done by means of a comprehensive discussion of one running example. Section 3.1 presents the dataset used for empirical analyses, while Sect. 3.2 discusses RRM-model estimation. Sections 3.3 and 3.4 discuss model fit, respectively the interpretation of estimation results, and Sect. 3.5 shows how estimated models can be used to forecast market shares (highlighting some of the RRM-model's important empirical properties). Section 3.6 concludes by discussing the out-of-sample validity of the RRM-model on the given dataset. Comparisons with the RUM-based MNL-model are provided throughout.

## 3.1 The Data

The data collection effort focused on route choice behavior among commuters who travel from home to work by car. A total of 550 people were sampled from an internet panel maintained by IntoMart, in April 2011.[1] Sampled individuals were at least 18-years old, owned a car, and were employed. It was taken care of that the sample was representative for the Dutch commuter in terms of gender, age and education level. Of these 550 people, 390 filled out the survey (implying a response rate of 71%).

Respondents to the survey were asked to imagine the hypothetical situation where they were planning a new commute from home to work (either because they had recently moved, or because their employer had recently moved, or because

[1] See Chorus and Bierlaire (under review) for an empirical comparison of several models (including a linear-additive RUM-model and an RRM-model) in the context of this dataset.

C. G. Chorus, *Random Regret-based Discrete Choice Modeling*,
SpringerBriefs in Business, DOI: 10.1007/978-3-642-29151-7_3,

| 1 | Route A | Route B | Route C |
|---|---|---|---|
| Average travel time (minutes) | 45 | 60 | 75 |
| Percentage of travel time in congestion (%) | 10% | 25% | 40% |
| Travel time variability (minutes) | ±5 | ±15 | ±25 |
| Travel costs (Euros) | €12,5 | €9 | €5,5 |
| YOUR CHOICE | ⍰ | ⍰ | ⍰ |

**Fig. 3.1** An example route choice-task

| Choice | TT1 | JAM1 | VAR1 | TC1 | TT2 | JAM2 | VAR2 | TC2 | TT3 | JAM3 | VAR3 | TC3 |
|---|---|---|---|---|---|---|---|---|---|---|---|---|
| 2 | 45 | 10 | 5 | 12.5 | 60 | 25 | 15 | 9 | 75 | 40 | 25 | 5.5 |
| 3 | 75 | 25 | 15 | 12.5 | 45 | 40 | 25 | 9 | 60 | 10 | 5 | 5.5 |
| 3 | 60 | 40 | 25 | 12.5 | 75 | 10 | 5 | 9 | 45 | 25 | 15 | 5.5 |
| 1 | 60 | 25 | 5 | 9 | 75 | 40 | 15 | 5.5 | 45 | 10 | 25 | 12.5 |
| ... | ... | ... | ... | ... | ... | ... | ... | ... | ... | ... | ... | ... |

**Fig. 3.2** Snapshot of the stated choice-data

they had started a new job). They were asked to choose between three different routes that differed in terms of the following four attributes, with three levels each: average door-to-door travel time (45, 60, 75 min), percentage of travel time spent in traffic jams (10, 25, 40%), travel time variability (±5, ±15, ±25 min), and total costs (€5.5, €9, €12.5). Using the Ngene-software package, a so-called 'optimal orthogonal in the differences'-design of choice sets was created to ensure a statistically efficient data collection. This design resulted in nine choice tasks per respondent (implying 3,510 choice observations in total). An example of a route-choice task is shown in Fig. 3.1.

Figure 3.2 shows a snapshot of the data that resulted from the Stated Choice-experiment introduced above. In this Figure, alternatives are numbered 1, 2 and 3 (rather than A, B and C); 'TT' stands for average travel time, 'JAM' stands for travel time spent in traffic jams, 'VAR' represents travel time variability and 'TC' denotes travel costs. Each row represents a choice made by an individual; column 'Choice' denotes the observed choice for each particular choice situation. For brevity, characteristics of the individual (age, education level, gender) are left out.

## 3.2 Model Estimation

Just like RUM-based MNL-models, RRM-based MNL-models can be estimated by means of Maximum Likelihood-routines. This means that in an iterative optimization process parameters are found that maximize the likelihood of the data, given the choice model and parameter estimates. In notation, this likelihood of the data as a function of a set of parameters inserted into a choice-model can be put as follows (Eq. 3.1):

$$L(\beta) = \prod_n \prod_i P_n(i|\beta)^{y_n(i)} \tag{3.1}$$

where $n$ denotes cases (choice observations), $P_n(i|\beta)$ gives the choice probability predicted by the choice model for alternative $i$ in case $n$ given $\beta$, and $y_n(i)$ equals 1 if alternative $i$ is chosen in case $n$, and 0 otherwise. For computationally pragmatic reasons, the natural logarithm of this likelihood-function (Eq. 3.2) is usually maximized rather than the likelihood itself:

$$LL(\beta) = \ln\left(\prod_n \prod_i P_n(i|\beta)^{y_n(i)}\right) = \sum_n \sum_i y_n(i) \cdot \ln(P_n(i|\beta)) \tag{3.2}$$

RRM-models can be estimated using two widely used software packages.[2] To start with, the model is incorporated into version 5.0 of the commercial NLOGIT-software (released early 2012). Using that software, RRM-models can be estimated by simply typing the command 'RRM' and selecting attributes to enter the choice model. In addition, RRM-models can be estimated using the discrete choice-software package Biogeme (Bierlaire 2003, 2008). Because this latter software package is freely available to scholars, students and practitioners, it is used in the remainder of this manual. Readers are referred to the Biogeme-manual (easily downloadable from the internet) for an in-depth introduction to this software-package. Figure 3.3 presents the input that is used to estimate an RRM-based MNL-model on the dataset described above. Again, alternatives are numbered 1, 2 and 3, and 'av1' is an availability indicator for alternative 1, etc. Like earlier, 'TT' stands for average travel time, etc. Finally, 'B' denotes an estimable parameter.

Users acquainted with Biogeme in the context of RUM-based models will note that this code resembles that of a RUM-based MNL-model. Note however, that the regret-function is entered under 'GeneralizedUtilities', rather than under 'Utilities'. This is needed to facilitate the use of ln- and exp-operators. Note also that minus-signs are needed to ensure that the *negative* of regret is maximized in the estimation process.

---

[2] Of course, the model can also be coded 'by hand' in packages like GAUSS or MATLAB.

```
[ModelDescription]

[Choice]
CHOICE

[Beta]
B_TT        0         -10000          10000        0
B_JAM       0         -10000          10000        0
B_VAR       0         -10000          10000        0
B_TC        0         -10000          10000        0

[Utilities]
1   Alt1   av1     $NONE
2   Alt2   av2     $NONE
3   Alt3   av3     $NONE

[GeneralizedUtilities]
1 - ln(1 + exp(B_TT * D12TT )) - ln(1 + exp(B_JAM * D12JAM ))
  - ln(1 + exp(B_VAR * D12VAR )) - ln(1 + exp(B_TC * D12TC ))
  - ln(1 + exp(B_TT * D13TT )) - ln(1 + exp(B_JAM * D13JAM ))
  - ln(1 + exp(B_VAR * D13VAR )) - ln(1 + exp(B_TC * D13TC ))

2 - ln(1 + exp(B_TT * D21TT )) - ln(1 + exp(B_JAM * D21JAM ))
  - ln(1 + exp(B_VAR * D21VAR )) - ln(1 + exp(B_TC * D21TC ))
  - ln(1 + exp(B_TT * D23TT )) - ln(1 + exp(B_JAM * D23JAM ))
  - ln(1 + exp(B_VAR * D23VAR )) - ln(1 + exp(B_TC * D23TC ))

3 - ln(1 + exp(B_TT * D31TT )) - ln(1 + exp(B_JAM * D31JAM ))
  - ln(1 + exp(B_VAR * D31VAR )) - ln(1 + exp(B_TC * D31TC ))
  - ln(1 + exp(B_TT * D32TT )) - ln(1 + exp(B_JAM * D32JAM ))
  - ln(1 + exp(B_VAR * D32VAR )) - ln(1 + exp(B_TC * D32TC ))

[Expressions]
D12TT = TT2 - TT1
...

av1 = 1
av2 = 1
av3 = 1

[Model]
$MNL
```

**Fig. 3.3** Biogeme-code for an RRM-based MNL-model

Although one may insert the actual attribute-values directly in the regret-function (under 'GeneralizedUtilities') this results in relatively long expressions that are more cumbersome to debug when applicable. It may be more convenient to create new variables (under 'Expressions') which give the differences in attribute-values that are needed in the regret-function. These new variables can then be inserted in the regret-function, leading to a more concise and manageable formulation. Figure 3.3 shows how these new variables are constructed, for the first attribute-difference: $D12TT = TT2-TT1$, being the travel time difference relevant for the comparison of alternative 1 with 2 (note that the difference that is relevant for the comparison of alternative 2 with 1 would read $D21TT = TT1-TT2$).

**Table 3.1** Estimation results for RUM-MNL and RRM-MNL models

| | RUM-MNL | | RRM-MNL | |
|---|---|---|---|---|
| | *Beta* | *t-value* | *Beta* | *t-value* |
| Average travel time (TT) | −0.0673 | −35.13 | −0.0468 | −32.50 |
| Percentage of travel time spent in traffic jams (JAM) | −0.0273 | −17.39 | −0.0181 | −16.66 |
| Travel time variability (VAR) | −0.0316 | −11.86 | −0.0210 | −11.86 |
| Travel costs (TC) | −0.173 | −21.52 | −0.113 | −20.28 |
| Number of observations | 3,510 | | 3,510 | |
| Null-loglikelihood | −3,856 | | −3,856 | |
| Final loglikelihood | −2,613 | | −2,605 | |
| Rho-square | 0.322 | | 0.324 | |

It is worth noting at this point that RRM-MNL models generally take longer to estimate than RUM-MNL models. The reason for this difference in speed is twofold: first, RRM-MNL models generally require a few more iteration steps before reaching convergence, although this difference tends to be small. Secondly, and more importantly, each single iteration step consumes more computation time for a RRM-MNL model than for its RUM-based counterpart. This is due to the simple fact that the regret-function, with its sequence of binary comparisons and its ln- and exp-operators, involves more (and more complicated) computations than a linear-additive utility function. Although the difference in runtimes between RRM and RUM is inconsequential in the context of MNL-models (which generally need seconds or at most minutes to converge), it may lead to more substantial time losses when more complicated Mixed Logit-models are estimated, especially when datasets are very large.

A number of observations can be made, based on the estimation results which are reported in Table 3.1. These observations relate to either model fit or parameter estimations and significance levels.

## 3.3 Model Fit

It appears that the RRM-MNL model has a slightly better fit with the data than its RUM-counterpart. The difference (8 Loglikelihood-points) is small but, when put to the Ben-Akiva and Swait test (1986) for nonnested models it turns out to be highly significant.[3] Notwithstanding this, a look at the models' rho-square shows

[3] The Ben-Akiva and Swait test gives an upper bound for the probability that, when some model A achieves a lower log-likelihood than some other (nonnested) model B, A is still the correct model of the data-generating process. This upper bound can therefore be considered a conservative proxy for the significance (or: p-value) of a difference in model fit between two nonnested models A and B.

that essentially the two models can be considered to fit the data equally well. As such, the results of the empirical example discussed here are in line with those obtained for other datasets. Roughly speaking, estimation results for a large and growing number of datasets (currently already well over 30) reveal that the RRM-MNL model outperforms its RUM-counterpart in about 50% of cases, the RUM-MNL model doing better on the other 50%. Differences in fit are generally small, but tend to be significant. See Chorus (2012) for a recent overview of empirical comparisons.[4] Note however that, as will be shown further below, the fact that differences in model fit are small does not imply that parameters, predictions and policy-implications derived from estimated models are similar across the two model types as well.

It is worth noting at this point that progress is being made to gain an understanding of what causes the RRM-MNL model to fit some datasets better than others (when compared to RUM). Although no definitive answers can be provided yet, empirical evidence based on sample-segmentation exercises seems to indicate that RRM fits choice-data better than RUM to the extent that participants are motivated and take the choice experiment seriously (in the case of Stated Choice data) and more generally to the extent that decision-makers consider it important to make the 'right' decision.[5] Respondents' motivation and their inclination to make the right decisions can be easily measured 'directly' using Likert-scale questions in the debriefing part of a Stated Choice experiment, for example involving questions like "How important was it for you to make the right decision in the hypothetical choice tasks presented to you during the experiment?". Alternatively, respondent motivation can be measured indirectly by means of the time taken by decision-makers to choose in the different choice tasks (RRM-models generally fit relatively well with choice-data generated by those individuals that have spent more time making decisions). Preliminary evidence also seems to suggest that RRM-MNL models generally perform somewhat worse when choice-situations include a so-called 'no-choice' or 'opt out' option, especially when that option cannot be characterized in terms of the attributes of the actual choice alternatives. See Chorus (2012) for a possible explanation for this preliminary finding.

---

[4] Published (or in press) papers have presented empirical applications of the RRM-model and comparisons with linear-additive RUM-models in the context of travelers' parking choices, information acquisition, shopping destination choices, mode-route choices (Chorus 2010), and departure time choices (Chorus and de Jong 2011); politicians' policy choices (Chorus et al. 2011); nature park visitors' site choices (Thiene et al. 2012; Boeri et al. in press); consumers' vehicle type choices (Hensher et al. in press); and dating website-visitors' choices among date-profiles (Chorus and Rose 2012). Studies comparing RRM and linear-additive RUM involving choices between leisure activities by senior citizens, and choices between medical treatments by patients, are underway.

[5] These empirical findings are in line with the more general notion, for which much empirical evidence exists, that the minimization of anticipated regret is an especially important determinant of choices when decision-makers feel that a choice is difficult and/or important (Zeelenberg and Pieters 2007).

## 3.4 Interpretation of Parameters

Whereas in a RUM-setting, a parameter estimate refers to the increase or decrease in utility associated with an alternative caused by a one-unit increase in an attribute's value, in an RRM-setting a parameter estimate refers to the *potential* (or: maximum) increase or decrease in regret associated with comparing a considered alternative with another alternative, caused by a one unit increase in an attribute's value. As will be illustrated further below, whether or not this potential level of regret is indeed attained by the increase in the attribute's value depends on the performance of the alternative in terms of the attribute, relative to the competition. In light of this conceptual difference between RUM- and RRM-parameters, a number of observations can be made in terms of the interpretation of parameters presented in Table 3.1:

First, as expected, parameters have the same sign in both models and estimates are of roughly the same order of magnitude across model types. However, these observations about magnitudes are not particularly interesting, nor do they necessarily hold in the context of other datasets. In fact, the size of RRM-estimates is inversely related to the size of the choice set. To see why this is the case, it should again be noted that the regret-function used in the RRM-model (Eq. 2.1) is the sum of all the strictly positive regrets that are generated by a series of binary comparisons with each of the other alternatives in the set in terms of every shared attribute. Therefore, if RRM-parameters would always be of the same size irrespective of the size of the choice set, this would imply the unrealistic scenario where in a large choice set, regret-levels and choice probabilities are more sensitive to changes in attribute-values than they are in a smaller choice set. A much more realistic scenario would be that the degree of sensitivity of regret-levels and choice probabilities to changes in attribute-values does not depend on the choice set size. This latter scenario implies that in the context of larger choice sets, smaller RRM-parameter estimates should be obtained and vice versa.

This conceptual expectation can be easily confirmed empirically: first, one estimates an RRM-model on actual or synthetic data. Then, one iteratively re-estimates the same model while, for each choice situation, randomly selecting an increasing number of non-chosen alternatives to be unavailable for choice (by doing so one gradually decreases the size of the choice set). The smaller the resulting choice set, the larger will be the estimated RRM-parameters.

As a result of this choice set size-sensitivity of RRM-parameters, direct comparisons with their RUM-counterparts are not particularly meaningful. Moreover, the choice set size-sensitivity of RRM-parameters implies that when RRM-estimates are smaller (larger) than RUM-estimates on a particular dataset, this does not mean to say that there is more (less) unobserved heterogeneity in the RRM-model than in the RUM-model, nor that particular attributes are less (more) important in the RRM-model than they are in the RUM-model.

Another consequence of this choice set size-dependency of RRM-estimates is that when an RRM-model is used for market share simulation purposes (see further below for an example), the choice set used for simulation should be of the same size as the one used for estimation in order to obtain unbiased predictions. When the size of the choice set that is used for forecasting is (much) larger than that of the choice set used for estimation, forecasted choice probabilities are biased towards *over* estimation of the sensitivity of choice probabilities to changes in attribute-values. Similarly, when the size of the choice set that is used for forecasting is (much) smaller than that of the choice set used for estimation, forecasted choice probabilities are biased towards *under* estimation of the sensitivity of choice probabilities to changes in attribute-values.[6]

One would expect the abovementioned choice set size-effect not to relate to *ratios of parameters* as the choice set size should affect every single parameter to the same extent. Again, this intuition can be easily confirmed[7] using synthetic data or actual choice data, by means of first estimating an RRM-model on the actual dataset and subsequently re-estimating the same model while, for each choice situation, randomly selecting one of the non-chosen alternatives to be unavailable for choice. Parameter *ratios* obtained in the context of the two estimation processes will be of roughly the same magnitude (one should of course allow for a small deviation which results from the randomness that is involved in the data generation process).

As a result of this choice set-insensitivity, parameter ratios in an RRM-model can be compared in a meaningful way with their RUM-counterparts: such a comparison gives a measurement of the differences across model types in terms of the relative importance of one attribute relative to that of another attribute.[8] Suppose that attributes $x$ and $y$ have associated parameters in the context of a RUM- and an RRM-model: $\beta_x^{RUM}$, $\beta_y^{RUM}$, $\beta_x^{RRM}$, $\beta_y^{RRM}$. Then the following result can be established in the context of a RUM-model: for a given alternative, an increase in $x$'s value of $\Delta_x$ units adds or removes as much utility to or from the alternative as does an increase in $y$'s value of $\Delta_x \cdot \beta_x^{RUM}/\beta_y^{RUM}$ units.[9] Likewise,

[6] A particularly fruitful direction for further research would be to explore—either empirically (by means of induction) or analytically (by means of deduction)—whether a mathematical function (or alternatively: a rule of thumb) exists that is able to map changes in choice set-size to changes in the size of RRM-based parameter estimates. Such a mapping or rule of thumb, if it exists, may then be used to correct parameters and eliminate biases in forecasted probabilities, when the forecast-choice set is of a different size than the estimation-choice set.

[7] This suggests that in order to attain unbiased ratios of parameter estimates in an RRM-MNL setting, one only needs to sample the chosen alternative and a small number of non-chosen alternatives.

[8] Of course, parameter ratios are not particularly suitable for assessing the relative importance of attributes within the context of one and the same estimated model, since parameter ratios are sensitive to the scale of measurement of the attributes. As such, the relative importance of for example travel time could be influenced by simply changing the scale from minutes to hours (which would imply a 60-fold increase of the magnitude of the associated parameter and parameter ratios). However, as long as the same scales are used across model types, parameter ratios can be meaningfully compared between models types, to highlight differences between model types in terms of the relative importance of attributes.

[9] In fact, when $y$ is a cost-attribute, $\beta_x^{RUM}/\beta_y^{RUM}$ gives the negative of the willingness-to-pay for a one unit change in attribute $x$. See Chap. 5 for a discussion of an RRM-based equivalent of this willingness-to-pay concept.

**Table 3.2** Comparison of relative importance of attributes (RUM- and RRM-parameter ratios)

| Number of units decrease in an attribute's value that is needed to create an as large increase in utility/potential decrease in regret as a 1 unit increase in travel cost | RUM | RRM | RRM/RUM[a] (%) |
|---|---|---|---|
| Travel time (minutes) | 2.6 | 2.4 | 94 |
| Percentage of travel time spent in traffic jams | 6.3 | 6.2 | 99 |
| Travel time variability (minutes) | 4.8 | 5.4 | 112 |

[a] Values lower than 100% indicate that the attribute is more important (relative to the cost attribute) in RRM than in RUM, and vice versa

in the context of an RRM-model, it holds that, for a given alternative, an increase in $x$'s value of $\Delta_x$ units has as much potential to generate or remove regret (associated with comparing the alternative with another alternative) as does an increase in $y$'s value of $\Delta_x \cdot \beta_x^{RRM}/\beta_y^{RRM}$ units. It is then easily understood that parameter ratios $\beta_x^{RUM}/\beta_y^{RUM}$ and $\beta_x^{RRM}/\beta_y^{RRM}$ provide an indication of relative importance of $x$ and $y$ in the context of the estimated RUM- and RRM-model, respectively.

To see how this type of comparison may be performed based on model estimation results, the relative importance of RUM- and RRM-parameters presented in Table 3.1 is provided in Table 3.2. The focus is on the situation where an alternative's travel cost is decreased by 1 €. This creates an increase in utility of 0.173 units, or a *potential* decrease in regret associated with comparing the alternative with another alternative of 0.113 units. Table 3.2 presents the equivalent number of units of decrease in other attributes that is needed to generate an as large increase in utility or potential decrease in regret. The table clearly shows that the relative importance of attributes as implied by the estimated RUM-, respectively RRM-models differ: the far right column shows that on these data the 'travel time'-attribute and the 'time spent in traffic'-attribute appear to be more important in the RRM-model (relative to the 'travel cost'-attribute) than in the RUM-model. For example: while the RRM-model predicts that a decrease of 2.4 min of travel time creates the same decrease in potential regret as does a 1 € decrease in travel costs, the RUM-model predicts that a 2.6 min decrease in travel time is needed to create the same increase in utility as does a 1 € decrease in costs. This implies that the relative importance of the travel time attribute (compared to the travel cost attribute) is smaller in a RUM-context than in a RRM-context (roughly 6% smaller, to be precise). On the other hand, it appears that the travel time variability-attribute is roughly 12% more important (relative to the travel cost-attribute) in the RUM-model than it is in the RRM-model.

Notwithstanding this comparability (across model types) of parameter ratios as measures of the relative importance of attributes, it should be kept in mind that the behavioral interpretation of parameter estimates themselves differs substantially between RUM- and RRM-based MNL-models, as mentioned above. This difference in parameter interpretation is best understood by means of going through a numerical example. Take the attribute travel cost, which—as shown in

Table 3.1—has a RUM-estimate of −0.173, and a RRM-estimate of −0.113. As said, the RUM-estimate implies that an increase of an alternative's travel costs of 1 € leads to a reduction of the utility of that alternative of size 0.173. In an RRM-setting, the increase in regret depends on the travel cost of a given alternative, relative to the travel cost of other alternatives. Take for example the choice set depicted in Fig. 3.1 and focus on route B which has a travel cost of 9 €. Increasing this travel cost to 10 € leads to an increase in regret associated with the bilateral comparison of B with A that equals $\ln(1 + \exp(-0.113*(12.5-10))) - \ln(1 + \exp(-0.113*(12.5-9))) = 0.047$. The same increase in travel cost leads to a larger increase in regret associated with the bilateral comparison of B with C (since $\ln(1 + \exp(-0.113*(5.5-10))) - \ln(1 + \exp(-0.113*(5.5-9))) = 0.069$). The total increase in regret caused by the increase in cost equals $0.047 + 0.069 = 0.116$ (note that the closeness of this total increase in regret to the absolute value of the estimate is coincidental).

This difference across bilateral comparisons in terms of the regret-increases caused by the increase in travel cost follows directly from the behavioral premises underlying the RRM-model: increasing the cost of an alternative (B) causes only a relatively small increase in regret associated with the comparison of that alternative with another alternative (A), when the other alternative is more expensive than the considered alternative and remains so after the travel cost increase. In both cases (before and after the increase in travel cost) the regret associated with comparing B with A is relatively small. On the other hand, increasing the travel cost of alternative B causes more regret associated with the comparison of that alternative with another alternative (C), when the other alternative was already cheaper before the change and becomes even more cheaper as a result of that change. As the initial difference in costs between B and C increases, the increase in regret caused by the additional 1 € increase in B's cost starts to approach 0.113. For example, an increase in B's cost from 50 to 51 € implies an increase in regret associated with the bilateral comparison of B with C equaling $\ln(1 + \exp(-0.113*(5.5-51))) - \ln(1 + \exp(-0.113*(5.5-50))) = 0.1123$.

Figure 3.4 provides a visual illustration of this argument: it plots the regret that is associated with comparing a considered alternative A with another alternative B in terms of some attribute $x$ on which B scores 0, and A's score is varied from −5 to +5, for a given parameter estimate of +1 (which implies that higher values are preferred over lower ones). In line with the abovementioned example, it is easily seen that a one-unit decrease in A's performance would generate almost no regret when the alternative performs much better than B before and after the change (e.g., a change from $x_A = 5$ to $x_A = 4$ as highlighted in the right hand side of Fig 3.4). Only when A already performs poorly on the attribute (relative to B) before the decrease in its attribute value, would an additional one unit decrease start to generate an amount of regret that approximates the parameter estimate (e.g., a change from $x_A = -4$ to $x_A = -5$ as highlighted in the left hand side of Fig 3.4).

To further illustrate the argument made above, Fig. 3.5 plots the increase in regret associated with comparing A with B in terms of $x$, that is caused by a one unit-decrease in $x_A$'s value.

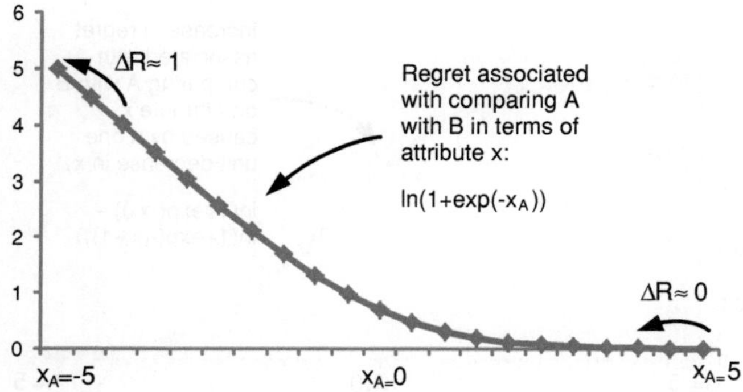


**Fig. 3.4** Regret associated with comparing A with B on attribute x ($\beta_x = 1$, $x_B = 0$)

The fact that (in isolation) RRM-estimates are not easily compared with RUM-estimates does not necessarily poses a problem—instead, one may argue that the RRM-estimates presents new and interesting behavioral interpretations that may be put to use for deriving new policy- and planning-implications. However, it is clear that for many purposes it is useful to be able to compare model estimation results directly across model types. A particularly strong candidate for enabling such a direct comparison, besides the notion of parameter ratios discussed above, is the concept of elasticity. An elasticity gives the percentage change in choice probability for an alternative per percentage change in one of its attributes' value. As a consequence of working with percentages rather than units, elasticities—in contrast with parameter ratios—are insensitive to the scale of the attribute which is generally considered an advantage. Although the mathematical derivation of RRM-based elasticities is outside the scope of this tutorial, the interested reader can find the derivation and an in-depth discussion (including an empirical example) in Hensher et al. (in press). See Train (2003) for an in-depth formal discussion and derivation of RUM-based elasticities.[10]

In both a RUM- and an RRM-setting, elasticities depend on the choice probability associated with an alternative in a particular choice situation. As such, they are alternative- and choice situation-specific. This is a result of the fact that the logit-model postulates (in line with intuition) that the largest impact of a percentage change in an attribute's value on an alternative's choice probability will be achieved when the alternative is a close competitor to one or more other alternatives in the set (i.e., when the alternatives have choice probabilities of similar magnitude). In contrast, when an alternative has a very low or a very high choice probability, changes in its attributes' values only have a small impact on the

[10] Note that the newest version of the NLOGIT-software package contains an automatic routine to compute RRM-elasticities, and that RRM-elasticities can also be generated by the newest version of Python-Biogeme.

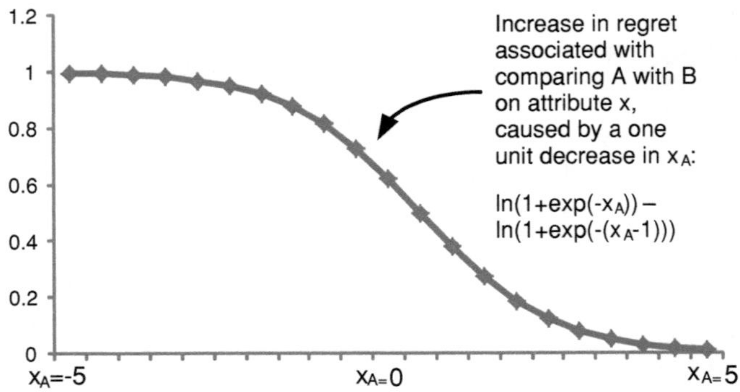


**Fig. 3.5** Increase in regret associated with comparing A with B on attribute x, caused by a one unit decrease in $x_A$'s value ($\beta_x = 1$, $x_B = 0$)

**Table 3.3** Comparison of relative importance of attributes (RUM- and RRM-elasticities)

| Percentage change in choice probability for an alternative per percentage change in an attribute's value | RUM | RRM | RRM/RUM[a] (%) |
|---|---|---|---|
| Travel time (minutes) | −2.82 | −3.03 | 107 |
| Percentage of travel time spent in traffic jams | −0.47 | −0.48 | 102 |
| Travel time variability (minutes) | −0.32 | −0.33 | 103 |
| Travel cost (euros) | −1.06 | −1.06 | 100 |

[a] In contrast with Table 3.2, here values *higher* than 100% indicate that the attribute is more important in RRM than in RUM, and vice versa

alternative's choice probability. This is why elasticities in a discrete choice-context are always computed for each alternative and for every choice situation, after which the average is taken to obtain a sample-wide estimate of elasticity.

In this respect, the notion of elasticity clearly contrasts with that of parameter ratios discussed further above, since the latter are not dependent on the choice probabilities of alternatives in particular choice situations. As such, one would not expect differences in elasticities between RRM- and RUM-based models to fully equal differences in parameter ratios. Nevertheless, it is instructive to show the RRM- and RUM-elasticities here (see Table 3.3) and briefly compare differences across model types with results presented in Table 3.2. Note that since the elasticity for the 'travel cost' attribute is the same for both the RRM- and RUM-model, the elasticity-ratios between model-types presented in the far-right column of Table 3.3 for the other attributes can be compared to the 'relative-to-cost' ratios between model types presented in the far-right column of Table 3.2.

The fact that for the attributes 'travel time' and 'time spent in traffic jams', RRM-elasticities are larger than their RUM-counterparts is in line with results obtained in the context of parameter ratios. The same holds for the fact that the difference in relative attribute importance between RUM and RRM is largest for the 'travel time' attribute. However, the result that the elasticity for the 'travel time variability'-attribute is larger in an RRM-than in a RUM-setting, contrasts with findings reported in the context of parameter-ratios. As mentioned above, differences like these can be expected based on the conceptual differences between these two measures of relative attribute importance.

## 3.5 Market Share Forecasting

Although the computation of choice probabilities or market shares based on estimated parameter values is just as straightforward for the RRM-MNL model as it is for its RUM-counterpart, some attention is paid to the topic in this tutorial. The reason is that by simulating market shares for hypothetical choice scenarios, two important properties of RRM-models can be highlighted effectively.

Consider a policy-maker or planner that wants to predict market shares for the three hypothetical routes presented in Fig. 3.1. the estimated RRM-model implies the following choice probabilities: P(A) = 67%, P(B) = 27%, P(C) = 6%. As illustration, part of the computations needed to obtain market shares is written out in full below.

$$
\begin{aligned}
R(A) = {} & LN(1 + EXP(-0{,}0468 * (60 - 45))) + LN(1 + EXP(-0{,}0468 * (75 - 45))) \\
& + LN(1 + EXP(-0{,}0181 * (25 - 10))) + LN(1 + EXP(-0{,}0181 * (40 - 10))) \\
& + LN(1 + EXP(-0{,}0210 * (15 - 5))) + LN(1 + EXP(-0{,}0210 * (25 - 5))) \\
& + LN(1 + EXP(-0{,}113 * (9 - 12.5))) + LN(1 + EXP(-0{,}113 * (5.5 - 12.5))) \\
= {} & 4.821
\end{aligned}
$$

$$R(B) = 5.734$$

$$R(C) = 7.185$$

$$
\begin{aligned}
P(A) &= \exp(-4.821) / (\exp(-4.281) + \exp(-5.734) + \exp(-7.185)) = 67\% \\
P(B) &= \exp(-5.734) / (\exp(-4.281) + \exp(-5.734) + \exp(-7.185)) = 27\% \\
P(C) &= \exp(-7.185) / (\exp(-4.281) + \exp(-5.734) + \exp(-7.185)) = 6\%
\end{aligned}
$$

Note that the estimated RUM-model implies the following choice probabilities: P(A) = 70%, P(B) = 23%, P(C) = 7%. It is easily seen that the differences in predicted market shares between RRM and RUM are non-trivial, and certainly more substantial than the very small difference in model fit. Note that empirical analyses on other datasets (not reported here), show that differences in predicted choice probabilities between estimated RRM-models and linear-additive RUM-models can become as large as 10 or more percentage points. Such double digit differences in market share forecasts are noteworthy and of significant practical importance.

In the remainder of this section, the above-mentioned example will be used to highlight two of RRM's most salient model properties: the presence of semi-compensatory decision-making, and the generation of a compromise effect. To start with the presence of semi-compensatory decision-making: it is instructive to note first that conventional RUM-models, like the linear-additive MNL-model form that is used throughout this tutorial, are so-called (fully) compensatory models. This means that the RUM-MNL model predicts that a deterioration of one attribute $x$ can be completely compensated (in terms of market share) by an equally large improvement of another, equally important attribute $y$. Furthermore, when the deteriorated attribute $x$ is twice as 'important' as the attribute $y$ that is improved (as would be implied by a twice as large parameter-estimate), a twice as large improvement of $y$ is needed to compensate, in terms of market share, for a given deterioration of attribute $x$.

Take for example the two attributes travel time (RUM-parameter $= -0.0673$/min) and travel costs (RUM-parameter $= -0.173$/€). By simulating market shares, it can be seen that 0.39 € are needed to compensate, in terms of market share, for a 1 min increase in travel time. Note that it is no coincidence that this value equals the ratio of parameters for time and cost ($0.39 = 0.0673/0.173$). Importantly, in a linear-additive RUM-model, needed levels of compensation do not depend on the initial relative performance of the alternatives in terms of the relevant attributes (travel time and cost). In other words, the obtained value of 0.39 € which is needed to compensate for a 1 min increase in travel time holds for each of the alternatives A, B and C in the choice situation depicted in Fig. 3.1, irrespective of their differences in performance in terms of travel time and costs.

In an RRM-MNL model things are quite different. To see this, let's compute in the context of the estimated RRM-MNL model the reduction in travel costs that is needed to offset a loss in market share caused by a 1 min increase in travel time. It will become clear that the needed level of compensation depends on how the alternative performs (in terms of time and costs) relative to the other alternatives in the set. Take for example alternative A, which is a fast but expensive route. A 1 min increase in travel time would require a 0.3 € reduction in travel costs to compensate for this loss in market share. Now take alternative C, which is a slow but cheap route. A 1 min increase in travel time would require a 0.5 € reduction in travel costs to compensate for this loss in market share. It is directly seen that this difference between the two alternatives, in terms of the level of compensation needed to offset a 1 min increase in travel time, is substantial. Yet, the difference is easily explained by building on the reasoning presented earlier in this chapter.

More specifically, it should be noted first that the 1 min increase in travel time of route A causes only a small increase in regret, since that route is by far the fastest one, and remains so after the marginal increase in travel time. In contrast, the alternative is a poor performer in terms of travel costs (relative to the other two alternatives), and as such a decrease in cost causes a relatively large decrease in regret. This implies that a (relatively) small cost decrease can compensate for the 1 min increase in travel time. For alternative C, the situation is exactly opposite: it is already the slowest route in the choice set, and as such a further increase in travel time causes a relatively large increase in regret. On the other hand, because it is the cheapest route, further reductions in travel costs result in only small reductions in regret. In combination, this implies that a relatively large reduction in travel cost is needed to compensate for the 1 min increase in travel time.

The above-mentioned difference is arguably the most important difference between RRM and linear-additive RUM in terms of predicted choice probabilities: in contrast with the linear-additive RUM-model, the RRM-model predicts that the level of improvement of one attribute that is needed to compensate (in terms of market share) a deterioration of another attribute depends on how the alternative performs, relative to the other alternatives, in terms of both attributes. This also implies that when some attribute ($x$) that is deteriorated is equally important as another one ($y$) that is improved (i.e., parameter estimates are the same), and the magnitude of the deterioration equals the magnitude of the improvement, the improvement in $y$ does not necessarily compensate for deterioration in $x$. As such, RRM is a so-called semi-compensatory model. As illustrated by the above example, the particular type of semi-compensatory nature exhibited by RRM-models makes that they can lead to markedly different policy implications when compared to linear-additive RUM-models, also when model fits and even parameter ratios are very similar across models.

The existence of so-called compromise effects, which is the second distinguishing RRM-model property that will be discussed in this chapter, follows directly from this particular brand of semi-compensatory behavior exhibited by RRM. The compromise effect, which is widely documented in empirical studies in consumer choice (e.g., Kivetz et al. 2004), states that there is a market share-bonus associated with being positioned 'in between' other available alternatives in terms of attribute-performance. In the choice situation depicted in Fig. 3.1, alternative B is such an 'in between' or 'compromise' alternative: while alternatives A and C have 'extreme' values for each of the attributes relative to the competition, alternative B always scores right in between: it is not the fastest route (which is A), but also not the slowest one (C); it is not the cheapest route (C), but also not the most expensive one (A), etc. when comparing market shares between model types, it becomes clear that the RRM-MNL model predicts a higher market share (27%) for this compromise alternative than does the RUM-MNL model (23%). This difference in predicted shares (4% points or roughly 17%) suggests that the RRM-model predicts a compromise effect (i.e., a market share bonus for the compromise alternative).

To see that this is indeed the case, the following line of argumentation may be used: as argued extensively in Sect. 3.3, the RRM-model predicts that having a (very) poor performance on one attribute causes much regret, while having a (very) strong performance on another attribute does not necessarily compensate for this (very) poor performance. As a result, it is more efficient (in terms of avoiding regret and as such gaining market share) to 'move to the center' of the choice set: an alternative that fails to have a really strong performance on any of the attributes (relative to the other alternatives) still only generates modest levels of regret as long as it does not have a particularly poor performance on any of the attributes either. This implies that alternative B, being a compromise alternative with a moderate performance on each attribute, has a relatively high market share in the context of an RRM-model (27%) when compared to a linear-additive RUM-model (23%).[11]

Note that the fact that RRM-models feature the compromise effect also implies that they do not exhibit the so-called IIA-property that characterizes linear-additive RUM-MNL models, even when errors are assumed to be i.i.d. (as is the case in the RRM-MNL model form). The IIA property says that the ratio of choice probabilities of any two alternatives is not affected by having a new alternative enter the choice set, nor by changing the performance of another alternative in the choice set. In contrast, RRM-models (including the MNL-model form) postulate that any ratio of choice-probabilities can be, and even is likely to be, affected by the presence of new alternatives or by changes in performance of any third alternative. For example, RRM-models postulate that when the attribute-levels of alternative B are changed to make alternative C a compromise alternative (i.e., positioned in between A and B), this leads to an increase of the choice probability ratio P(C)/P(A).

It is important to note here that by not exhibiting the IIA-property, RRM-MNL models do not yet deal with the type of IIA-property commonly referred to as the 'red bus-blue bus problem'. This latter type of IIA—from which RRM's MNL also suffers, just like RUM's linear-additive MNL-model—originates from ignored correlations between the error terms of competing alternatives, and can be addressed in an RRM-framework in a similar way as in a RUM-framework: by creating nests in Nested or Mixed Logit model-extensions.

## 3.6 Out-of-Sample Validity

In many of the fields that use discrete choice experiments, it is considered good practice to test a model's predictive validity on a part of the dataset that was not used for estimation purposes. When comparing the performance of two or more models in the context of a particular dataset, this out-of-sample validity is usually considered an equally important criterion as model fit. There are several metrics available to test a model's out-of-sample validity, of which two particularly popular ones are used here to compare the estimated RUM- and RRM-models: the log-likelihood on the validation sample, and the hit rate. To this aim, the data were split into an estimation-sample and a validation-sample by means of randomly selecting two thirds of cases for estimation and leaving the remaining one third (1,192 cases) for testing out-of-sample predictive ability. Estimation results of the RUM- and RRM-models on the estimation-sample are very similar to those

[11] See Chorus and Bierlaire (under review) for an in-depth analysis of models that are able to capture compromise effects (including the RRM-model), based on these data.

**Table 3.4** Out-of-sample predictive ability of RUM- and RRM-models

| | RUM-model | RRM-model |
|---|---|---|
| Log-likelihood of validation sample | −913 | −913 |
| Hit rate | 778/1,192 = 65.3% | 786/1,192 = 65.9% |

reported in Table 3.1 (in terms of parameter values as well as model fit statistics) and are not reported here.

First, the log-likelihood of the validation sample is reported for each of the two models. This log-likelihood is computed by applying Eq. (3.2) on the *validation* sample, using the parameters estimated on the *estimation* sample. In other words, this metric gives an indication of how likely the data in the *validation* sample are, given the model as estimated on the *estimation* sample. Higher (less negative) values of the log-likelihood are preferred over lower ones. Table 3.4 shows that the linear-additive RUM-model and the RRM-model perform equally well in terms of this metric, each of them scoring a log-likelihood value of −913.

As a second way to test the models' out-of-sample predictive ability, so-called hit rates are computed for each model. These are computed by first predicting, for each case in the *validation* set, the choice probabilities for every alternative using the RUM- and RRM-models as estimated on the *estimation* set (see the previous section for an explanation of how these choice probabilities or 'market shares' are computed). Subsequently, for each case in the validation sample and for both the RUM- and the RRM-model, the alternative that has the highest predicted choice probability is compared with the actually chosen alternative. If the two alternatives (predicted and observed) coincide, a score of one is obtained for that case, and otherwise a score of zero. Higher hit rates are of course preferred over lower ones. Results are as follows: the linear-additive RUM-model has a hit rate of 778/1,192 = 65.3%, while the RRM-model achieves a hit rate of 786/1,192 = 65.9%. Although differences are fairly small, it does appear that the RRM-model slightly outperforms the linear-additive MNL-model when it comes to identifying the most popular alternative in the choice sets.

To conclude, it should be noted here that—as was the case with model fit comparisons—the relative out-of-sample performance of RUM- and RRM-models differs across datasets. As of now, there is no conclusive evidence of one model performing consistently better than another model in terms of, for example, out-of-sample log-likelihood or hit rate (nor is such evidence to be expected). It is found, moreover, that differences in model fit between RUM- and RRM-models are not necessarily good predictors for the direction of differences in out-of-sample predictive ability; i.e., many situations occur where one model does better in terms of model fit, while the other model does better in terms of one or more out-of-sample predictive ability metrics. This is actually to be expected, given the probabilistic nature of discrete choice models.

## References

Ben-Akiva ME, Swait J (1986) The Akaike likelihood ratio index. Transp Sci 20:133–136

Bierlaire M (2003) Biogeme: a free package for the estimation of discrete choice models. In: Proceedings of the 3rd Swiss transportation research conference, Ascona, Switzerland

Bierlaire M (2008) An introduction to BIOGEME Version 1.7, biogeme.epfl.ch

Boeri M, Longo A, Doherty E, Hynes S (in press) Site choices in recreational demand: a matter of utility maximization or regret minimization? J Environ Econ Policy doi:10.1080/21606544.2011.640844

Chorus CG (2010) A new model of random regret minimization. Eur J Transp Infrastruct 10:181–196

Chorus CG (2012) Random regret minimization: an overview of model properties and empirical evidence. Transp Rev 32:75–92

Chorus CG, Bierlaire M (under review) An empirical comparison of travel choice models that generate compromise effects

Chorus CG, de Jong GC (2011) Modeling experienced accessibility for utility-maximizers and regret-minimizers. J Transp Geogr 19:1155–1162

Chorus CG, Rose JK (2012) Selecting a date: A matter of regret and compromises. In: Hess S, Daly A (eds) Choice modelling: the state of the art and the state of practice: selected papers from the second international choice modelling conference, Edward Elgar, UK

Chorus CG, Annema JA, Mouter N, van Wee GP (2011) Modeling politicians' preferences for road pricing policies: a regret-based and utilitarian perspective. Transp Policy 18:856–861

Hensher DA, Greene W, Chorus CG (in press) Random regret minimisation or random utility maximisation: an exploratory analysis in the context of automobile fuel choice. J Adv Transp. doi:10.1002/atr.188

Kivetz R, Netzer O, Srinivasan V (2004) Alternative models for capturing the compromise effect. J Mark Res 41:237–257

Thiene M, Boeri M, Chorus CG (2012) Random regret minimization: exploration of a new choice model for environmental and resource economics. Environ Resour Econ 51:413–429

Train KE (2003) Discrete choice methods with simulations. Cambridge University Press, Cambridge

Zeelenberg M, Pieters R (2007) A theory of regret regulation 1.0. J Consum Psychol 17:3–18

# Chapter 4
# Applicability of Random Regret Minimization-Models, and Their Strong and Weak Points

**Abstract** This chapter starts (Sect. 4.1) with discussing the applicability of the RRM-model in other than the quite straightforward choice context presented in Chap. 3. More specifically, it is explained how RRM can be applied to model non-generic attributes, constants, non-continuous attributes, and interactions with socio-demographic variables. Subsequently, Sects. 4.2 and 4.3 present strong, respectively weak points of the RRM-model.

## 4.1 General Applicability

The empirical example presented in Chap. 3 was based on a choice situation that is particularly intuitive to explain the working of RRM-models, in the sense that it featured a choice situation where alternatives were presented in terms of generic attributes (i.e., attributes that are shared by all alternatives) that were also continuous. The notion of regret is particularly easy to understand and model in such a context, as illustrated in that Chapter. However, the RRM-approach can also be readily applied in choice situations where one or more attributes are non-generic and/or non-continuous; in these situations, a number of approaches can be chosen.

### *4.1.1 Non-Generic Attributes*

Sometimes, the analyst might want to specify one or more attributes in a way that they are specific to only one or a subset of the alternatives in the choice set: take for example the situation where a choice set contains car and train alternatives and a researcher wants to estimate travel time-parameters for each of these two modes

C. G. Chorus, *Random Regret-based Discrete Choice Modeling*,
SpringerBriefs in Business, DOI: 10.1007/978-3-642-29151-7_4,

separately (i.e., he or she wants to estimate $\beta_{TT}^{car}$, $\beta_{TT}^{train}$), to account for the fact that train-travel time may be valued differently than car-travel time. In such a situation, the RRM-model can be applied by simply constructing the following function for, for example, the regret associated with comparing a car option with a train option in terms of their travel times: $R_{TT}^{car} = \ln(1 + \exp[\beta_{TT}^{train} \cdot TT^{train} - \beta_{TT}^{car} \cdot TT^{car}])$. When considering that $\ln(1 + \exp[\beta_m \cdot (x_{jm} - x_{im})]) = \ln(1 + \exp[(\beta_m \cdot x_{jm} - \beta_m \cdot x_{im})])$ it is easily seen that the implementation of non-generic attributes merely results in a special case of the more general formulation of the regret-model presented in Eq. (2.1). In a more general sense, the regret associated with an alternative $i$ in a situation where all attributes $m$ are alternative-specific can be written as follows: $R_i = \sum_{j \neq i} \sum_m \ln(1 + \exp[\beta_m^j \cdot x_{jm} - \beta_m^i \cdot x_{im}])$. Usually, a mixture of generic and alternative-specific attributes may be present, leading to a combination of the above equation and the conventional regret-function given in Eq. (2.1).

A different, slightly more involved approach consists of assuming that all car-alternatives feature zero minutes of train travel time, and that train alternatives feature zero minutes of car travel time. The regret-function for a car option (more specifically, the part of the regret-function that refers to comparing the car option with a train option in terms of their travel times) would then be written as follows: $R_{TT}^{car} = \ln(1 + \exp[\beta_{TT}^{car} \cdot (0 - TT^{car})]) + \ln(1 + \exp[\beta_{TT}^{train} \cdot (TT^{train} - 0)])$.[1]

### 4.1.2 Constants

A special case of a non-generic attribute is the alternative-specific constant. Suppose that there are three alternatives in the choice set, which are intrinsically unique and 'labeled' in the sense that alternative A is always a car-alternative and alternative B is always a bus alternative while alternative C is always a metro alternative. Constants for two of these three travel mode types (say, $\beta^{car}$ and $\beta^{bus}$) can then be identified while a third is normalized to zero, just like in a linear-additive RUM-model. These constants can be modeled in an RRM-model by constructing the following function for, for example, the regret associated with comparing a car option with its competitors in terms of their 'labels': $R^{car} = \ln(1 + \exp[\beta^{bus} - \beta^{car}]) + \ln(1 + \exp[\beta^{metro} - \beta^{car}])$, which in light of the normalization of $\beta^{metro}$ can be simplified to $\ln(1 + exp[\beta^{bus} - \beta^{car}]) + \ln(1 + exp[-\beta^{car}])$. In a more general sense, the regret associated with an alternative $i$ in a situation where every alternative has a constant $\beta_0$ (one of which needs to be normalized to zero) can be written as follows: $R_i = \sum_{j \neq i} \ln(1 + exp[\beta_0^j - \beta_0^i]) + \sum_{j \neq i} \sum_m (1 + exp[\beta_m \cdot (x_{jm} - x_{im})])$.

[1] This function can of course be simplified: $R_{TT}^{car} = \ln(1 + \exp[-\beta_{TT}^{car} \cdot TT^{car}]) + \ln(1 + \exp[\beta_{TT}^{train} \cdot TT^{train}])$.

### *4.1.3 Non-continuous (Categorical) Attributes*

Next is the issue of the implementation of non-continuous attributes: take for example the situation where a choice set contains a number of train options, which differ in terms of—among other attributes—the availability of wireless internet. More specifically, suppose that the internet-availability attribute can take on the value 'no internet', 'internet for a 1 euro fee' and 'internet for free'. In this case, the attribute is clearly non-continuous and there is no single parameter that is likely to describe the preference of travelers for the entire attribute-range. This kind of variables can be dealt with in an RRM-model as follows: first, like is the case in a linear-additive RUM-model, two dummy variables (0/1-variables) are created to represent the three-category non-continuous attribute internet-availability', and whose associated parameters will be estimated. More specifically: each category is represented by a dummy, minus one category (parameter) which needs to be normalized (to zero) to ensure identification.

Suppose that 'no internet' is considered the base level, implying that an 'internet for a 1 euro fee'-dummy ($x_{fee}$) and an 'internet for free'-dummy ($x_{free}$) are created for each alternative. This in turn implies that parameters $\beta_{fee}$ and $\beta_{free}$ are estimated. Assume a particular trinary choice-situation where alternative A is of type 'no access', and that B is of type 'access for a 1 euro fee', while C is of type 'free access'.[2] Now, just like would the case in a linear-additive RUM-setting, the following dummies are created:$x^A_{fee} = x^A_{free} = 0$; $x^B_{fee} = 1$, $x^B_{free} = 0$; $x^C_{fee} = 0$, $x^C_{free} = 1$. The part of the regret-function of alternative A ('no access') that relates to comparing the alternative with, for example, alternative B ('access for a 1 euro fee') in terms of internet-availability then can be written as follows: $R^A_{int} = \ln\left(1+\exp\left[\beta^B_{fee}\cdot x^B_{fee}+\beta^B_{free}\cdot x^B_{free}-\left(\beta^A_{fee}\cdot x^A_{fee}+\beta^A_{free}\cdot x^A_{free}\right)\right]\right)$, which for this particular case can be simplified to $R^A_{int} = \ln\left(1+\exp\left(\beta^B_{fee}\right)\right)$. In the above formulation the term '$\beta^B_{fee}\cdot x^B_{fee}+\beta^B_{free}\cdot x^B_{free}$' refers to the performance of B in terms of internet-availability (higher values lead to *more* regret for A) and the term '$\beta^A_{fee}\cdot x^A_{fee}+\beta^A_{free}\cdot x^A_{free}$' refers to the performance of A in terms of internet-availability (higher values lead to *less* regret for A). Regrets associated with comparing A with C, B with A, B with C, C with A and C with B are derived in a similar fashion.

[2] It is important to note here that if type A would *always* be of type 'no access', etc., this would imply that the alternatives are *de facto* labeled and that the approach outlined in Sect. 4.1.2 should be used instead: preferences for types are then treated as constants. The text in Sect. 4.1.3 refers to the situation where A's type (as well as that of B and C) varies across cases in the dataset.

### 4.1.4 Hybrid RUM-RRM Models

Notwithstanding that the approaches outlined above provide effective means to deal with non-generic variables (including constants) and/or non-continuous variables in regret-functions, it is instructive to note that these kinds of variables can also be treated 'outside' the regret-function, as utilitarian variables. More generally speaking, one may choose to position one or more attributes outside the regret-function while having others enter the regret-function. Behaviorally, such a hybrid model implies that some attributes are assumed by the researcher to be processed by the decision-maker in a linear-additive RUM-way, while others are assumed to be processed in a RRM-fashion. If there are $Q$ attributes being implemented outside the regret-function, and $M - Q$ inside the regret-function, the systematic part of such a hybrid utility/regret-function would look like this (Eq. 4.1):

$$V_i = \sum_{m=1..Q} \beta_m \cdot x_{im} - \sum_{j \neq i} \sum_{m=Q+1..M} \ln\left(1 + \exp\left[\beta_m \cdot \left(x_{jm} - x_{im}\right)\right]\right) \qquad (4.1)$$

Choice probabilities are given by the utility-based MNL formulation given in Eq. (2.3). Preliminary empirical analyses Chorus et al. (in press) show that for approximately one out of every three datasets, a hybrid model-specification achieves a higher model fit and/or out-of-sample validity than a model that assumes only RUM-heuristics or only RRM-heuristics for each and every attribute.

### 4.1.5 Interactions with Socio-Demographic Variables

Besides being able to deal with non-generic and non-continuous variables as shown above, the RRM-model also readily applies to situations where individual-level or socio-demographic variables are interacted with constants and/or attribute-level parameters. More specifically, this is done in almost the exact same way as in linear-additive RUM: take for example a parameter referring to a price-attribute ($\beta_{price}$), which is interacted with the individual-level variable '*age*' to represent the fact that different age-segments may have different tastes for that attribute. This is simply done by replacing in the all regret-functions $\beta_{price}$ by $\beta_{price} + \beta_{price}^{age} \cdot age$. Parameter $\beta_{price}^{age}$ then refers to the additional regret that is caused by a particular price-difference between two alternatives, for a one-year age increase. A similar formulation holds for interactions with constants. Hensher et al. (in press) provide an empirical example of an RRM-model including socio-demographic interactions.

## 4.2 RRM: Strong Points

It is increasingly acknowledged that the RRM-model in a number of ways provides an appealing alternative to conventional RUM-models. First, its regret-based premises provide a behaviorally intuitive alternative to the utility-maximization paradigm that has dominated the field of choice modeling and related fields for decades. Many feel that it pays off (in terms of insights gained) to study discrete choice-behavior from different behavioral perspectives, and the RRM-model provides an intuitive and formally tractable way of doing so. The RRM-model 'tells a story' that is fundamentally different from the linear-additive RUM-storyline, and as such it may lead to new behavioral insights and new policy- and planning-implications. In this regard an important advantage of RRM is that it offers policy-makers and demand managers an instrument to identify and capitalize on the way the composition of the traveler's (or: consumer's) choice set influences behavior. In other words, RRM offers a quantitative tool to help policy-makers and planners with 'choice set-engineering'.

Furthermore, there is ample evidence from all corners of the social sciences [see Zeelenberg and Pieters (2007) for an overview] that the anticipation of regret is an important determinant of choice behavior; the RRM paradigm provides an opportunity to translate this notion into an operational discrete choice-modeling perspective. In addition, two of RRM's most important properties (semi-compensatory decision-making and the existence of a compromise-effect) have been well-established empirically in numerous studies. In a RRM-model context, these properties readily emerge from the model's underlying behavioral premises.

In addition, the RRM-model appears to do well empirically: on many datasets (around 50% of those studies so far) RRM-models achieve a better model fit and/or out-of-sample validity than their RUM-counterparts. At the aggregate level of the dataset, differences between RRM and linear-additive RUM in terms of model fit and out-of-sample validity are generally small. However, estimation results and derived quantities—such as parameter ratios, elasticities and simulated choice probabilities for particular policy scenarios—may differ quite substantially across models. As such, the RRM-approach presents a new behavioral model perspective that has a considerable policy-relevance.

When compared to other so-called non-RUM models, RRM-models have the advantage of being very parsimonious (they consume no additional parameters compared to RUM's most basic linear-additive model form), while at the same time remaining quite user-friendly and compatible with (or: incorporated in) some of the field's most-used software packages. As such, perhaps more than many other non-RUM models, the RRM-model is a viable option also for those modelers that lack the time or econometric experience to code and apply highly complex choice models.

## 4.3 RRM: Weak Points

Having mentioned these strong points, it is clear that the RRM-model has its drawbacks and limitations as well. The issue of run-times was mentioned earlier in this tutorial: especially in the context of Mixed Logit models and very large datasets, RRM-models take considerably more time to converge than their

RUM-counterparts. The author's experience suggests that RRM run-times may become twice or three times as high as RUM run-times, depending on the error term specification in Mixed Logit models, the number of attributes, and the choice set size. For some applications, this might limit the applicability of RRM-models. Note however, that in the context of most MNL-models, run-times are not an issue at all. A related issue is that of combinatorial explosion: since RRM's regret-functions involve comparisons with each alternative in the choice set on every attribute, there is a risk of combinatorial explosion: run-times grow faster than linearly as a function of choice set size. Although most choice sets encountered in discrete choice studies deal with choice sets of small or moderate size, some studies (e.g., some destination choice studies) involve very large choice sets of hundreds or even thousands of alternatives, which would lead to prohibitively high run-times for RRM-models.[3]

More generally speaking, the RRM-MNL model—while being more 'simple' than most other non-RUM models—is still slightly more complex than RUM's workhorse, the linear additive MNL-model. As is shown in Chap. 3, the difference in complexity between the two paradigms is hardly noticeable when dealing with small choice sets, small numbers of attributes and simple specifications involving no interactions with socio-demographic variables. However, for the analysis of more involved choice situations, the RRM-model can become somewhat more cumbersome to code and interpret than the RUM-MNL model.

Finally, the fact that the RRM-model is relatively new—which is an appealing characteristic for some modelers, especially in academia—may prove a disadvantage as well, for example when it comes to communicating results to policy-makers and planners. Furthermore, it is obvious that the RRM-model is not yet as well understood as its counterpart, the linear-additive RUM-MNL model. For example, it is not yet completely clear how the RRM-model can be used for economic appraisal, although progress is being made in this regard (e.g., Chorus and de Jong 2011; Chorus 2012; Chorus in press—see also Sects. 5.3 and 5.4 in the next Chapter). This clearly contrasts with linear-additive RUM-models' level of applicability for economic appraisal, which has been well-documented over the past few decades. The topic of economic appraisal is just one example of the still considerable knowledge-gap that exists between RRM and linear-additive RUM-models.

However, considering the amount of attention currently being devoted to the RRM-model by a growing number of choice-modelers it may be expected that in time this knowledge gap will diminish, making the RRM-model an increasingly competitive alternative to linear-additive RUM-models. Take for example Bekhor et al. (in press), who derive and apply an RRM-based Stochastic User

[3] However, it may be noted that the behavioral realism of the assumption that decision-makers would actually consider dozens, let alone hundreds, of alternatives when choosing is increasingly debated in fields like travel behavior as well as consumer psychology. As a result, more and more researchers call for the use of small choice sets in choice model estimation and application, even when in practice the universal set of alternatives may be very large.

Equilibrium (as a counterpart of the conventional RUM-based Stochastic User Equilibrium) which enables the regret-based prediction of equilibrium traffic flows on transport networks. The next Chapter discusses a number of research areas where recent attempts have been made to further increase the scope and usefulness of RRM.

## References

Bekhor S, Chorus CG, Toledo T (in press) A stochastic user equilibrium formulation for the random regret minimization-based route choice model. Transp Res Rec

Chorus CG (2012) Random regret minimization: an overview of model properties and empirical evidence. Transp Rev 32:75–92

Chorus CG (in press) Logsums for utility-maximizers and regret-minimizers, and their relation with desirability and satisfaction. Transp Res A

Chorus CG, de Jong GC (2011) Modeling experienced accessibility for utility-maximizers and regret-minimizers. J Transp Geogr 19:1155–1162

Chorus CG, Rose JM, Hensher DA (under review) Regret minimization or utility maximization: it depends on the attribute

Hensher DA, Greene W, Chorus CG (in press) Random regret minimisation or random utility maximisation: an exploratory analysis in the context of automobile fuel choice. J Adv Transp. doi:10.1002/atr.188

Zeelenberg M, Pieters R (2007) A theory of regret regulation 1.0. J Consum Psychol 17:3–18

# Chapter 5
# Selection of Recent Developments in RRM-Modeling

**Abstract** Very recently, a number of developments have taken place that have further broadened the scope of the Random Regret Minimization approach to discrete choice modeling. This chapter provides brief introductions to some of these developments, without aiming to be complete with respect to the number of developments covered nor the amount of detail provided for any one particular development. Instead, this chapter's sections aim to provide some first insights into these recent developments. Wherever relevant, references are provided to research papers providing more elaborate discussions. Avenues for further research are discussed at the end of each section. Finally note that these recent developments are discussed here in the context of transportation-related choices. However, relevance and implications for other fields can be easily seen. Sections 5.1, 5.3, and 5.4 draw from recently published papers or papers that are currently in press. Citations are provided wherever relevant.

## 5.1 Random Regret Minimization for Decision-Making Under Uncertainty

The term 'uncertainty' here refers to situations where decision-makers, at the time they make a decision, are not certain about the values of (some of) the attributes of alternatives under consideration.[1] Take for example a traveler that chooses between different routes, but cannot assess with full certainty what travel times he or she may encounter during the trip. Therefore, the choice between routes is generally a choice under uncertainty, and the traveler is then assumed to take into account this uncertainty when making decisions.

[1] This section draws from Chorus (2012).



C. G. Chorus, *Random Regret-based Discrete Choice Modeling*,
SpringerBriefs in Business, DOI: 10.1007/978-3-642-29151-7_5,

Extending the RRM-approach towards the context of analyzing decisions made under uncertainty is relatively straightforward, and similar to the extension of (Random) Utility-Theory to (Random) Expected Utility-Theory. Assume that a traveler conceives uncertainty in terms of the possibility of occurrence of different states of the world *s*, where each state has a (perceived) probability of occurrence. These probabilities are represented by a probability density function[2] *f*(*s*), although also discrete representations of uncertainty are possible. The result of this uncertainty is that attribute-values and hence also regret become state-specific and that the traveler is now assumed to choose based on the minimization of *expected* regret *E*(*R*), where the expectation is taken with respect to *f*(*s*). In some states of the world, a choice for alternative *i* will generate high levels of regret, whereas in other states, it will not—the decision-maker is assumed to weigh these regrets according to the probability of occurrence of different states, and is assumed to choose the alternative with minimum expected regret. In notation (Eq. 5.1):

$$E(R_i) = \int_s \left[ \sum_{j \neq i} \sum_m \ln\left(1 + \exp\left[\beta_m \cdot \left(x_{jm}^s - x_{im}^s\right)\right]\right) \cdot f(s) \right] ds \qquad (5.1)$$

The same MNL-type choice probabilities as formulated before apply,[3] expect that $R_i$ and $R_j$ are replaced by $E(R_i)$ and $E(R_j)$ respectively (Eq. 5.2):

$$P(i) = \frac{\exp(-E(R_i))}{\sum_{j=1..J} \exp(-E(R_j))} \qquad (5.2)$$

Some readers may note that, when written in this expected regret-form, the RRM-model reminds of two popular theories of decision making under uncertainty: Regret Theory (RT—Loomes and Sugden 1982) and Prospect Theory (PT—Kahneman and Tversky 1979). Notwithstanding these connections at first sight between (Expected) RRM on the one hand and RT and PT on the other hand, it can be argued that (Expected) RRM differs from each of these theories in terms of a number of important aspects.

First, RT focuses on decision-making under uncertainty only and is designed with the aim to capture choice-anomalies that are not being dealt with in neoclassical Expected Utility-theory, like preference reversals and common ratio effects. In contrast, while RRM may be extended towards the analysis of decision-making under uncertainty as shown above, it is primarily developed for the analysis of choice under conditions of full certainty, where it captures semi-compensatory choice behavior and choice set-composition effects. Second, while RT is focused on the study of single-attribute choices often used in the field of

---

[2] Formally speaking, the use of a probability density function suggests that the term 'risk' would be more appropriate than the term 'uncertainty'. However, for ease of communication the latter term is used here.

[3] See Chorus (under review) for a recent empirical application (estimation) of this Expected RRM-model.

behavioral and experimental economics (mostly involving monetary gambles), RRM is explicitly designed to model choices between multi-attribute alternatives like those presented in Stated Choice-experiments often used in the field of, for example, travel demand and consumer choice analysis. Thirdly, where RT postulates that regret arises by comparing the utilities of *alternatives*, RRM postulates that regret by comparing alternatives in terms of each of their *attributes*: the essence of RRM lies in its hypothesis that when making multi-attribute choices, regret arises from the likely situation where one has to put up with a relatively poor performance on one or more attributes to arrive at a relatively strong performance on other attributes. These three conceptual differences between RRM and RT in combination translate into substantial differences in formal regret-functions, predictions and scope/applicability. As a result, it makes more sense conceptually to position RRM-theory as a regret-based counterpart of linear-additive RUM-theory, rather than a discrete choice-counterpart of Regret Theory.

The connection between RRM and Prospect Theory (PT) follows from the fact that both postulate that the evaluation of an alternative depends on its performance relative to a reference point, and that losses compared to that reference point loom larger than gains of the same magnitude. But also here, the differences between the two choice-theories are larger than the seemingly unifying concept of reference-dependency would suggest. Firstly, RRM's reference points are given by the attribute-level performance of foregone alternatives, whereas PT's reference point is generally given by some operationalization of the status quo (or: current wealth). As a result, whereas PT postulates that preferences are *context*-dependent, RRM claims that they are *choice set*-dependent. Secondly, RRM captures loss aversion in a different way than does PT, beyond the difference in adopted reference point. More specifically, RRM assumes that the consideration of gains does not matter at all while PT assumes that perceived gains are still a substantial determinant of choice behavior. Furthermore, the RRM-model features no additional parameter(s), like PT does, to measure the degree of loss aversion. Thirdly, whereas PT is explicitly positioned as a model of decision-making under uncertainty (featuring a probability-weighting function to represent potential probability-distortions), RRM is tailored for choice-situations where attribute-values are known to the decision-maker and as such it features no behavioral postulates concerning how people deal with probabilities beyond the use of a probability density function as outlined above. In sum: although at first sight RRM's use of a reference point may suggest a deep connection between RRM and PT, this connection is not that intuitive and persuasive when more profoundly inspecting the two theories of choice.

An important avenue for further research is to empirically test how RRM (using the notion of expected regret) compares to linear-additive RUM (using the notion of expected utility) in the context of decision-making under uncertainty. In addition, empirical comparisons with non-Expected Utility models such as Prospect Theory-based models are needed to determine how each of these decision-making models performs empirically.

## 5.2 RRM for Modeling Household or Group Decision-Making

Many choices are made a group or household level, rather than by one individual in isolation. Take for example the situation where a household decides where to go on vacation. In a RUM-setting, this household decision making process has been traditionally modeled by assuming a household utility function which represents the aggregated utilities of all members. It is assumed that the household trades off the utilities assigned by different household members to various choice options, using weights that represent the 'importance' or 'bargaining power' of different members. Recent work suggests that the RRM-framework can provide a particularly intuitive alternative modeling framework for these household (or: group) decision-making processes. The focus is here on the modeling framework at a conceptual level, rather than on data-issues and model estimation processes.

In notation: denote choice-alternatives by $i$, $j$. Denote individuals by $z$ (and assume there are $Z$ individuals in the household). Denote the regret that an individual member of the household $z$ attaches to an alternative $i$ by $R_i^z$. Denote the importance, or bargaining power, of individual $z$ within the household by $\alpha_z$. It may then be postulated that, when choosing between alternatives, households aim to minimize (anticipated) random household regret, and that the level of random household regret that is associated with a considered alternative $i$ is composed of an i.i.d. random error, which represents unobserved heterogeneity in household regret and whose negative is Extreme Value Type I-distributed, and a systematic household regret $HR_i$. The core of this systematic household regret is the following term (Eq. 5.3):

$$HR_{i\leftrightarrow j}^z = \ln\left(1 + \exp\left[\alpha_z \cdot \left(R_i^z - R_j^z\right)\right]\right) \tag{5.3}$$

$HR_{i\leftrightarrow j}^z$ gives the amount of household regret that is associated with comparing the regrets that are associated with alternative $i$ and $j$ respectively, as anticipated by member $z$. Here, $R_i^z = \sum_{j \neq i} \sum_m \ln\left(1 + \exp[\beta_m^z \cdot (x_{jm}^z - x_{im}^z)]\right)$ gives the amount of regret that is associated with alternative $i$, as anticipated by member $z$. If $i$ has a lower regret than $j$ in the eye of this particular household member, then the comparison of $R_i^z$ with $R_j^z$ generates (almost) zero household regret. When $i$ has a higher regret than $j$ in the eye of the household member, then the comparison of $R_i^z$ with $R_j^z$ generates a level of household regret that grows as a semi-linear function of the importance of the household member and the difference between $R_i^z$ and $R_j^z$ (in the eye of that member). Note that the indexation of $x_{jm}^z$, ${}_{im}^{z}{}^{z}$ and $\beta_m^z$ with respect to $z$ implies that perceptions and preferences of members with respect to the attributes of alternatives may differ. Note furthermore, that the above formulation does not specify random errors at the level of the regret anticipated by individual household members. However, these can be added and accommodated by using an error component Mixed Logit formulation at the household member level.[4]

The term $HR^z_{i\leftrightarrow j}$ discussed above is computed for each household member, and for all binary comparisons between the considered alternative $i$ and every other alternative $j \neq i$. Systematic over-all household regret is then the sum of all these terms (Eq. 5.4):

$$HR_i = \sum_{j\neq i}\sum_{z=1..Z} HR^z_{i\leftrightarrow j} = \sum_{j\neq i}\sum_{z=1..Z} \ln\left(1 + \exp\left[\alpha_z \cdot \left(R^z_i - R^z_j\right)\right]\right) \tag{5.4}$$

The same MNL-type choice-probabilities as given in Eq. (2.4) apply. The difference between this household regret-function and the individual-level regret-function discussed earlier in this tutorial can be put as follows: the individual-level regret-function that is discussed throughout this tutorial refers to weighing the performance of different alternatives in terms of their attributes. Weights represent the importance of attributes. In contrast, the household level formulation of the regret-function discussed in this section refers to weighing the performance of different alternatives in terms of the regret (or: utility) they generate for particular household members. Weights represent the importance or bargaining power of those members. The household regret-function incorporates the individual level regret-function to represent the preferences of its members, although this individual level regret-function may also be replaced by an individual level utility function.

Conceptually, the difference between an RRM and a linear-additive RUM approach to modeling group decision making is that the former assumes that a group aims to minimize regret at a group level, and as such is likely to choose compromise alternatives which are not extremely disliked by any of the household members. In contrast, using a compensatory decision rule at the group level like that of linear-additive RUM-models would presuppose that the dislike of one household member can be readily compensated for by the satisfaction of another member. Intuitively, one would expect that many group decision making processes result in compromises to make sure that no particular group member is deeply unsatisfied with the outcome of the decision-making process. As such the RRM-approach is likely to be a particularly effective and intuitive modeling paradigm for such group decision-making processes. However, a very obvious direction for further research is to empirically compare the performance of the RRM-based group decision making model with that of the mainstream linear-additive RUM-based approach. Work is underway to perform such a test, using data on household decision-making in the context of vehicle purchase.

---

[4] At this point it may also be noted that rather than assuming regret-minimization calculus at both the level of the household and the level of the household member, one may assume that while a *household* minimizes regret, every *member* of the household maximizes a linear-additive utility function. In that case the term $R^z_i - R^z_j$ in Eq. (5.3) becomes $V^z_j - V^z_i$, and these utilities can be written as in Eq. (2.2), with indexation for the household member.

## 5.3 The Expected Regret of a Choice-Situation: An RRM-based Logsum

It is well known that the welfare economic impacts of (transport) policies can be rigorously analyzed using the notion of the expected utility associated with a (traveler's) choice set.[5] Furthermore, it has been shown (e.g., Small and Rosen 1981; Ben-Akiva and Lerman 1985) that, in the context of RUM-based MNL models, the expected utility associated with a choice set takes the form of a so-called Logsum-measure ($LS_{RUM}$, Eq. (5.5), using the notation provided in Chap. 2):

$$LS_{RUM} = E\left[\max_{j=1..J}\{U_j\}\right] = \int\limits_{\varepsilon} \left[\max_{j=1..J}\{U_j\} \cdot f(\varepsilon)\right] d\varepsilon = \ln\left[\sum_{j=1..J} \exp\left[V_j\right]\right] \tag{5.5}$$

Recently, it has been shown in Chorus (in press) that an RRM-based counterpart of this RUM-Logsum exists (denoted $LS_{RRM}$ from here on) and that it has an intuitive interpretation as the expected regret that the traveler receives from the choice among the alternatives. More specifically, Chorus (in press) shows that this amount of expected regret is derived as follows (Eq. 5.6)[6]:

$$\begin{aligned} LS_{RRM} &= E\left[\min_{j=1..J}\{RR_j\}\right] = \int\limits_{\varepsilon} \left[\min_{j=1..J}\{RR_j\} \cdot f(\varepsilon)\right] d\varepsilon \\ &= -\ln\left[\sum_{j=1..J} \exp\left[-R_j\right]\right] \end{aligned} \tag{5.6}$$

There are a number of important differences between the RUM-based Logsum and its RRM-based counterpart. First note that the RUM-paradigm postulates that improving the performance of one of the alternatives in the choice set will always increase the expected utility (or: RUM-Logsum) of the choice set. The RRM-Logsum does not exhibit this property, as is easily seen by considering the following illustrative example (depicted in Fig. 5.1): suppose a traveler faces a choice between three alternatives with two attributes each: A = {2,1}, B = {$x_B$, 1.5} and C = {1,2}. That is, A performs well on the first attribute ($x$) but less well on the second attribute ($y$). The opposite is the case for alternative C, while B has an in-between performance on the second attribute. Its first attribute-value is varied. Assume both attributes have unit weight, and that higher values are preferred over lower ones. Figure 5.1 shows that regret is relatively low for very low values of $x_B$

[5] This section draws from Chorus (in press).

[6] Note that of course, as a planner or policy-maker one would aim at minimizing the value of this RRM-Logsum, whereas in a RUM-setting one would aim at maximizing the RUM-Logsum.

**Fig. 5.1** The RRM-Logsum (*solid line*) as a function of attribute performance [*source* Chorus (in press)]

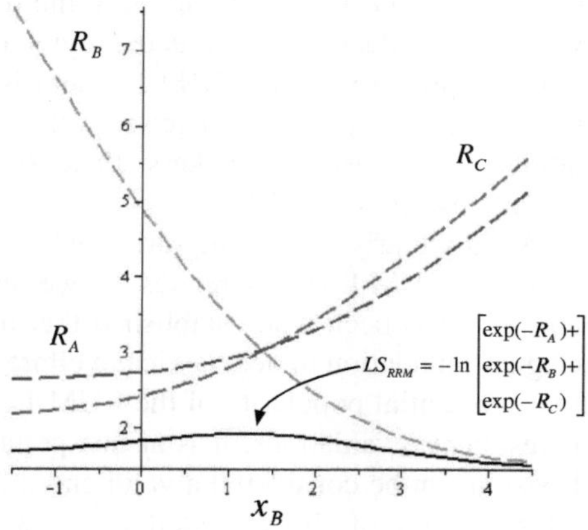


and that it reaches its lowest point for very high values of the attribute. For average values of $x_B$ however, regret is relatively high. This implies that improving the attribute when it performs (very) poorly only decreases the expected regret associated with the choice situation when the improvement is such that the alternative attains a (very) strong performance on the attribute.

The behavioral intuition behind this non-monotonicity of the RRM-Logsum can be put as follows: when alternative B has a very poor performance on attribute $x$, this makes that there is a very high level of regret associated with that alternative. However, this very high level of regret for the *alternative* does not translate into a high level of expected regret for the *choice situation*, because there are alternatives with much lower regrets available. Furthermore, the fact that B performs poorly on $x$ implies that alternatives A and C have relatively low regrets (since comparisons with B barely generate regret at all).

When $x_B$ attains a more intermediate value, B's regret decreases but the regrets associated with the other two alternatives increase since now, the comparison of A and C with B does generate some regret. The result is a situation where all regrets are of roughly similar magnitude, and that the expected regret associated with the choice situation *increases*—because the regret associated with the alternative(s) with lowest regrets (A, C) increases. Only when $x_B$ becomes very large, does the expected regret associated with the choice situation start to decline: in that situation, the regret associated with A and C grows, but since B's regret is now (much) lower than these two regrets this increase in A's and B's regret becomes more and more irrelevant. The fact that B's regret declines therefore finally starts resulting in a lower level of expected regret associated with the choice situation as a whole. Only when $x_B$ becomes very high, a lower level of expected regret is obtained than was associated with the situation when B performed very poorly on $x$.

In general, in can be said that a change in the choice set that makes the traveler's decision more difficult—in the sense that due to the change there is no clear 'winner' among the alternatives—will increase the expected regret associated with the choice situation. Changes that do create clear winners, even when this comes at the cost of a deteriorating the performance of one or more alternatives, will lead to a decrease in expected regret associated with the choice situation. Note that this property of the RRM-Logsum is in fact in line with consumer research that suggests that larger choice sets without clear 'winners' may be considered less attractive by decision-makers than small sets containing a clear 'winner' (e.g., Schwartz et al. 2002).

A particularly interesting direction for further research would be to investigate how the RRM-Logsum relates to welfare economic principles. As mentioned above, it has been well-established that the RUM-Logsum has a strong and very elegant foundation in neoclassical welfare theory. This rigorous connection has led to a substantial popularity of the RUM-Logsum in transport project (infrastructure investment) evaluations. It is at this point not yet clear to what extent the RRM-Logsum can be considered a valid and useful welfare measure (beyond its natural interpretation of the expected regret associated with a choice situation), and establishing to what extent it is (in-)consistent with welfare economics may be considered a potentially very fruitful research direction.

## 5.4 An RRM-based Formulation of 'Value of Time'

It is very well known that in a linear-additive RUM-MNL model context involving travel choices, the ratio of partial derivatives of the utility function with respect to the travel time attribute (denoted TT) and the travel cost attribute (TC) has an intuitive interpretation as the Value of Time (VoT from here on).[7] This VoT gives the amount of money that a traveler is willing to pay for a one unit *decrease* in travel time.[8] As such, the notion of VoT is crucial input in quantitative evaluations of transport policies such as policies aimed at decreasing congestion (travel times) in traffic networks. In the context of linear-additive RUM-MNL models, this ratio of partial derivatives results in a very easy-to-compute and interpret formulation (Eq. 5.7):

$$VoT_{RUM} = \frac{\partial V_i}{\partial TT_i} \Big/ \frac{\partial V_i}{\partial TC_i} = \frac{\beta_{TT}}{\beta_{TC}} \tag{5.7}$$

In other words, VoT in a linear-additive RUM-MNL context is given by the ratio of the parameters estimated for the travel time and travel cost-attributes. The RRM-counterpart of this ratio of partial derivatives takes on a quite different form (Eq. 5.8):

---

[7] This sections draws from Chorus (2012).

[8] More in general, the ratio of partial derivatives of a utility function with respect to one of its attributes and the cost attribute gives the (negative of the) willingness to pay for a one unit increase in the non-cost attribute.

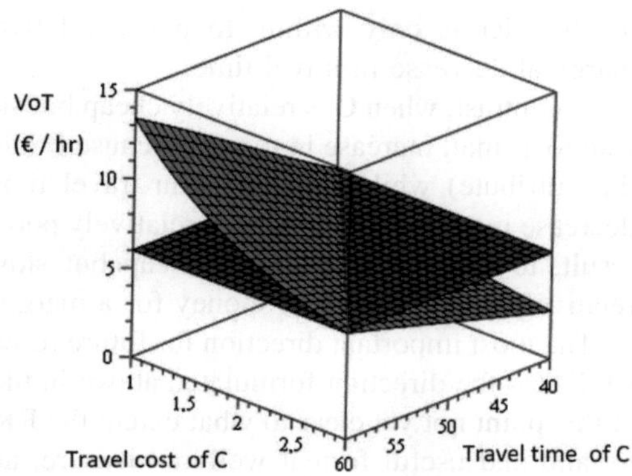


**Fig. 5.2** Value of Time for RUM (*horizontal plane*), RRM (*diagonal plane*) [*source* Chorus (2012)]

$$VoT_{RRM} = \frac{\partial R_i}{\partial TT_i} / \frac{\partial R_i}{\partial TC_i} = \frac{\sum_{j \neq i} \left( \frac{-\beta_{TT}}{1 + \frac{1}{\exp[\beta_{TT} \cdot (TT_j - TT_i)]}} \right)}{\sum_{j \neq i} \left( \frac{-\beta_{TC}}{1 + \frac{1}{\exp[\beta_{TC} \cdot (TC_j - TC_i)]}} \right)} \tag{5.8}$$

The crucial difference between the RUM- and the RRM-formulation of VoT, is that the latter results in alternative- and choice set-specific measures of VoT. That is, the current travel times and travel costs of both the considered alternative and its competitors enter the VoT-equation, which implies that RRM-based VoT-measures will generally change when the choice set changes in terms of alternatives' performance on times and costs. As the numerical example presented directly below will show, these changes are fully in line with the properties of the RRM-model—more specifically the particular type of semi-compensatory behavior it exhibits.

Assume the following choice situation: a traveler faces three routes A, B and C, which are evaluated in terms of their travel times ($\beta_{TT} = -0.1$/min) and costs ($\beta_{TC} = -1$/€) only. Route A is 'cheap but slow' ($TT = 60$, $TC = 1$), while B is fast but expensive ($TT = 40$, $TC = 3$). Figure 5.2 plots alternative C's VoT as given by a linear-additive RUM-based MNL-model (horizontal plane) and its RRM-counterpart (diagonal plane), as a function of the alternative's travel time and cost. Of course, linear-additive RUM-VoT is not affected by changes in C's time and cost, and always equals 6 €/h. In contrast, RRM-VoT is a function of C's performance, relative to that of the other alternatives in the set: when C is relatively cheap and slow, RRM-VoT can become as high as 14 €/h, while dropping to 3 €/h when C is relatively fast but expensive.

These results are in line with RRM's specific brand of semi-compensatory behavior highlighted earlier in this section: when C is relatively fast but expensive, an increase in travel cost would lead to much additional regret (because C already performs relatively poorly on that attribute) while a decrease in travel time would only lead to a small decrease in regret (since C already has a relatively strong performance on the attribute). As a result, to the extent that C is fast but expensive, the traveler is only willing to pay a relatively small amount of money for a marginal decrease in travel time.

In contrast, when C is relatively cheap but slow, an increase in travel cost would lead to a small increase in regret (because C has a relatively strong performance on that attribute) while a decrease in travel time would lead to a relatively large decrease in regret (since C has a relatively poor performance on the attribute). As a result, to the extent that C is cheap but slow, the traveler is willing to pay a relatively large amount of money for a marginal decrease in travel time.

The most important direction for future research into RRM-VoTs would be very similar to the direction formulated above in the context of the RRM-Logsum: it is at this point not yet clear to what extent the RRM-VoT measure can be considered a valid and useful formal welfare measure, and establishing to what extent it is (in-) consistent with welfare economics may be considered a potentially very fruitful research direction.

## References

Ben-Akiva ME, Lerman SR (1985) Discrete choice analysis: theory and application to travel demand. The MIT Press, Cambridge

Chorus CG (2012) Random regret minimization: an overview of model properties and empirical evidence. Transp Rev 32:75–92

Chorus CG (under review) Valuating ex-post travel information: a matter of balancing regrets

Chorus CG (in press) Logsums for utility-maximizers and regret-minimizers, and their relation with desirability and satisfaction. Transp Res A

Kahneman D, Tversky A (1979) Prospect theory: an analysis of decision under risk. Econometrica 47:263–291

Loomes G, Sugden R (1982) Regret theory: an alternative theory of rational choice under uncertainty. Econ J 92:805–824

Schwartz B, Ward A, Monterosso J, Lyubomirsky S, White K, Lehman DR (2002) Maximizing versus satisficing: happiness is a matter of choice. J Pers Soc Psychol 83:1178–1197

Small KA, Rosen HS (1981) Applied welfare economics with discrete choice models. Econometrica 49:105–130

# Appendix
# Investigating the Assumption of i.i.d. Errors in RRM-Models

Like RUM-based MNL-models, RRM's MNL-model assumes that random errors are independent (uncorrelated) across alternatives. This assumption, which is a crucial condition for obtaining closed form choice probabilities, may appear counterintuitive at first sight because of the fact that an alternative's observed regret is a function of the performance of the alternative relative to that of all other alternatives in the choice set. One may ask: how can the *random* component of a considered alternative's regret be safely assumed to be independent from other alternatives' random regrets, when the *observed* regret of the considered alternative is a function of the performance of all alternatives in terms of observed attributes? More specifically, one may be inclined to expect that if one alternative's unobserved regret decreases, the unobserved regret of a randomly selected competing alternative must increase. In this Appendix, an argumentation is provided for the use of independent (or: uncorrelated) errors in RRM-models.

Before investigating to what extent the assumption of independent errors in RRM-models is appropriate, it is important to clearly state what these errors actually are assumed to represent. It is generally understood that random errors in a choice model aim to represent unobserved heterogeneity in regret (or, for that matter, utility) that is the result of a combination of one or more of the following three factors: (i) *measurement error*[1] from the side of the analyst in terms of the decision-makers' perceptions and tastes; (ii) *omission of relevant attributes* by the analyst; (iii) *random behavior* by decision-makers. It is directly seen that this latter source—randomness in behavior—does not lead to correlations between random errors of competing alternatives in RRM-models. Below, each of the two remaining sources of random errors will be discussed in terms of their potential contribution to correlations between random errors in RRM models.

---

[1] Alternatively, one may refer to this category as representing *unobserved heterogeneity* in perceptions and tastes of decision-makers.

C. G. Chorus, *Random Regret-based Discrete Choice Modeling*,
SpringerBriefs in Business, DOI: 10.1007/978-3-642-29151-7,

Regarding source (i), consider the situation where the perception[2] of one of the decision-makers concerning one of the attributes of one of the alternatives—say, alternative A—is incorrectly measured by the analyst. It can be easily verified that in the context of an RRM-model this causes a negative covariation between the random error associated with alternative A and that of a randomly selected competing alternative B: when higher values of the attribute are preferred over lower ones, over estimation[3] by the analyst of A's attribute-value causes an under estimation of A's regret and therefore an *increase* in the random error associated with A; simultaneously it causes an over estimation of B's regret and therefore a *decrease* in the random error associated with B. However, it is also easily seen that this covariation is of limited magnitude. This is a direct result of the fact that the attribute-regret-function has a convex, rather than a linear, shape (see Figs. 2.1 and 3.4 for visualizations). This convexity implies that the extent to which over estimation of A's attribute-value causes an increase in A's random error does not equal the extent to which it causes a decrease in B's random error. For example, suppose without loss of general applicability that—under the condition that alternative A's attribute is incorrectly measured—the alternative performs less well than B in terms of that attribute. In that case, correcting the over estimation of A's performance on that attribute (which is what the random error in fact does) will lead to a relatively large increase in A's regret and a relatively small decrease in B's regret.

Given that the sign and magnitude of measurement errors will vary across cases (decision-makers), these measurement errors will induce a negative correlation between random errors of alternatives at the aggregate level of the dataset.[4] The size of this correlation will depend on the number of measurement errors and there magnitudes.

Regarding the omission of attributes—source (ii)—consider the situation where a decision-maker considers a particular attribute as relevant and includes it in his or her regret-function, whereas the analyst ignores the attribute. Interestingly and perhaps counter intuitively at first sight, it turns out that irrespective of how each of the alternatives in the choice set performs in terms of the attribute and irrespective of the (sign and magnitude of the) associated taste of the individual, ignoring the attribute results in an *under estimation* of the regret for each alternative in the set. Therefore, omitting the attribute results in an *increase* in the

[2] It can be easily verified that a similar line of argumentation holds for measurement error (or: unobserved heterogeneity) regarding *tastes* of decision-makers.

[3] It is easily seen that a similar argument applies to the situation where lower values of the attribute are preferred over higher ones, and for the situation where the attribute's value is under estimated, rather than over estimated.

[4] Note that when the attribute-value is also measured incorrectly for alternative B this may lead to either a partial or full cancellation of the correlation (when the measurement errors have the same sign, for example when the performance of both alternatives in terms of the attribute is over estimated) or to an amplification of the covariance (when the signs of the measurement errors differ between A and B).

random errors that are associated with each alternative. The reason for this lies in the fact that by definition, an alternative's regret consists of an additive function of attribute-regrets, which themselves are strictly positive (see Eq. 2.1). As a result, there will be a positive covariation between all random errors in the individual's choice set, due to the omission of the attribute by the analyst. However, the magnitude of this covariation will be limited, since obviously the level of 'ignored' or omitted regret will be smaller for alternatives that perform relatively well in terms of the omitted attribute, than for alternatives with a relatively poor performance on that attribute.

As long as there is variation across decision-makers in terms of whether they take into account the omitted attribute in their individual regret-function, omitting a relevant attribute will therefore induce a positive correlation between random errors of alternatives at the aggregate level of the dataset. The size of this correlation will depend on the number of omitted attributes.

The above discussion clearly indicates that while measurement errors result in *negative* correlations between random errors in RRM-models, the omission of attributes that are relevant for a subset of decision-makers results in *positive* correlations. In both cases, correlations are not perfect (i.e., they differ from –1 and 1, respectively). Of course, this does not yet give an indication of the level of overall correlation between random errors in RRM-models since the relative importance of each of the two sources is not known and as such it is not known to what extent the two correlations cancel each other out. Although it seems unlikely that the two sources of correlation will completely cancel each other out, it also seems unlikely that one of the two sources will strongly dominate (causing positive or negative correlations in random errors). Furthermore, it should be noted that the third source of random error (random behavior by decision-makers) generates i.i.d. errors and as such will dampen the effect of any correlation resulting from the combination of correlations caused by sources (i) and (ii).

In the end it seems likely that the extent to which random errors in RRM-models can be safely assumed to be independent, will depend to a large extent on the data used for model estimation. This leads to an important final remark: the question to what extent the assumption of independent errors is realistic in the context of RRM-models remains largely an empirical one. That is, if an RRM-model assumes independent errors while the corresponding data generation process (i.e., observed choice behavior) in combination with the analyst's measurement and modeling process involves negatively (or: positively) correlated errors, this will result in a relatively low fit between the estimated model and the data. In other words, to the extent that an RRM-model with independent errors (like the RRM-MNL model) achieves a good model fit on a particular dataset, this can be considered evidence that the assumption of independent errors was in fact justified in the context of that dataset.

MIX
Papier aus verantwortungsvollen Quellen
Paper from responsible sources
FSC® C105338

If you have any concerns about our products,
you can contact us on
**ProductSafety@springernature.com**

In case Publisher is established outside the EU,
the EU authorized representative is:
**Springer Nature Customer Service Center GmbH**
**Europaplatz 3, 69115 Heidelberg, Germany**

Printed by Libri Plureos GmbH
in Hamburg, Germany